Almost America

Also by Steve Tally

*Bland Ambition: From Adams to Quayle—The Cranks,
Criminals, Tax Cheats, and Golfers Who Made It
to Vice President*

Almost America

From the Colonists to Clinton:
A "What If" History of the U.S.

Steve Tally

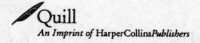
Quill
An Imprint of HarperCollinsPublishers

HarperCollins books may be purchased for educational, business, or
sales promotional use. For information please write: Special Markets
Department, HarperCollins Publishers, Inc., 10 East 53rd Street, New
York, NY 10022.

FIRST EDITION

Designed by Elina D. Nudelman

Library of Congress Cataloging-in-Publication Data

Tally, Steve.
    Almost America : from the colonists to Clinton : a "what if" history of
the U.S. / Steve Tally. — 1st ed.
    p. cm.
    Includes index
    ISBN 0-380-80091-8
    1. United States — History — Miscelleanea. I. Title.

E179.T35 2000
973—dc21
                                                        00-055296

00 01 02 03 04 RRD 10 9 8 7 6 5 4 3 2

Contents

The New Lights effected a great and revolutionary spiritual revival in the colonies. What if their preachers had been cowed by the government that opposed them?

Humiliated after losing New York to the British, the Americans were forced to retreat. What if George Washington and his troops had not crossed the Delaware River to fight the British again?

In a battle that was little noticed at the time, and is hardly remembered today, George Rogers Clark defeated the British on the western frontier and captured Fort Sackville at the small village of Vincennes. What if Clark had failed?

In the early years of the republic, the government of the United States was nothing but a "cobweb confederation." What if the government had continued under the Articles of Confederation, despite their many faults?

Lt. Col. Robert E. Lee, the hero of Mexico City and Harper's Ferry, was offered the command of the U.S. Army days before being offered the command of the rebel forces. What if he had accepted Lincoln's proposal?

The Confederate Army of Northern Virginia, defeated at Gettysburg, had to retreat through Pennsylvania and Maryland to reach safety on the opposite side of the Mason-Dixon line. What if Union general George Meade had counterattacked and cut off Lee's retreat?

Vice President Andrew Johnson and other top government officials were supposed to have been killed on the night Abraham Lincoln was assassinated. What if the conspiracy had been successful?

Democrat Samuel Tilden received three hundred thousand more votes than Republican opponent Rutherford B. Hayes in the election of 1876, and lost. What if Samuel Tilden had become president in 1877?

Andrew Carnegie sold his steel company to J. P. Morgan in 1901, despite having misgivings about the way the deal was financed. What if Carnegie had refused to sell?

What if Teddy Roosevelt had followed through on his threat to abolish the nation's most popular collegiate sport?

Author's Note

I have been sent on a fool's errand, and many nights I have lain awake and wondered what it was in my résumé that made my editor decide that I was the right person for this job.

Counterposing historical facts and speculative vignettes, this book attempts to show what our history might have been like had particular events taken a different turn.

The question "What if?" is probably as old as history itself. Our good friend Mark Twain, not himself a historian but rather a man for the ages, once expressed a view on such revisionist history. He visited the house of Abraham Lincoln with a congressman who remarked, "What a pity it was that fate did not intend that Lincoln should marry Ann Rutledge. It seems that fate governs our lives and plans history in advance."

"Yes," replied Twain, "had Lincoln married the dear one of his heart's love he might have led a happy but obscure life and the world would never have heard of him. Happiness seeks obscurity to enjoy itself. A good-looking milkmaid might have kept Alexander the Great from conquering the world."

"Well, doesn't that prove that what is to be will be?" asked the congressman.

Twain shook his head. "The only thing it proves is that what has been was."

What-might-have-been history is known among academics and other learned people as counterfactual history, and when they use the term they are usually not being complimentary. Among the hundreds of books that helped in the preparation of this book, one historian referred to the "forbidden delights that such counterfactual musings can provide to any suitably degenerate mind." Well.

Historians dislike counterfactual history for the simple reason that too many people don't know the basics of history already. The reasonable fear is that, because of a lack of knowledge about what really happened, the reader will become befuddled and forget what is real and what is fiction. Unlike these scholars of little faith, I trust you to be discerning enough to tell the difference.

Given historians' opposition, you might well wonder why any-one would be interested in writing counterfactual history. Well, the truth is, the most important reason is that it is fun to think about what might have been. But there is a nutritious, good-for-you reason, too. Like a golfer who lines up a putt from the opposite side of the cup, sometimes the best way to understand something is to look at it from a different direction. Counterfactual history allows us to pick up a moment in history, roll it around in our hands a bit, and try to come away with new knowledge of what was happening at the time. Coun-terexamples are always a useful technique in explanatory writing.

In this book there are a few self-imposed ground rules. To the extent that I thought it possible, I tried to make the counterfactual scenarios plausible. Adolf Hitler doesn't step into a time machine to join Robert E. Lee at Gettysburg, at least not in this book. The sec-ond ground rule is that the counterfactuals are based on the decisions of human beings, not on acts of God. It would be interesting to think about what might have happened in the Vietnam War if JFK had lived after the shooting in Dallas in 1963, but that wasn't going to happen without divine intervention. Focusing on decisions allows us to second-guess those decisions, and second-guessing is always good sport.

If you care to second-guess my second guesses, you may. Simply go to this book's Web site at *www.AlmostAmerica.com* and join in a dis-cussion.

Acknowledgments

THIS is the hardest page of the book to write. Not because I'm afraid I'll leave someone out—I'm fairly certain I'll do that—but because a few words cannot do justice to the people who truly deserve my appreciation.

The first person I need to thank is my wife, Lisa Hunt Tally. For those of us who are not full-time authors, writing a book is an interruption in our lives and the lives of our families, and I appreciate the sacrifices that my family members have made on my behalf.

The book would have been impossible without my research assistant, J. Jacob Jones, who became used to my desperate phone calls asking him to help me find out what city some minor member of Congress was in on a particular day more than one hundred years ago. Portions of this book are by design speculative, but he helped me make sure that the facts written through the speculations are correct.

And I want to thank my agent, Madeleine Morel, and editor, Stephen S. Power, for initiating the project, and editors Bret Witter, Sarah Durand, Susan Andrews, and Jennifer Hershey for bringing the book to fruition.

Chapter 1

The Slumbering American Spirit

❧

The New Lights effected a great and revolutionary spiritual revival in the colonies. What if their preachers had been cowed by the government that opposed them?

IN the late summer of 1775, American commander Benedict Arnold and his small ragtag army of rebels prepared for battle by following a bizarre ritual. The troops and their chaplain assembled in a Presbyterian church in Northampton, Massachusetts. Following a sermon, the chaplain, Arnold, and his officers stepped behind the pulpit and lowered themselves into the brick tomb of a minister, George Whitefield, who had died five years before.

Standing in the tomb, the chaplain removed the lid of the mahogany-colored coffin, revealing Whitefield's skeleton. The chaplain then cut up small pieces of the collar and cuffs of Whitefield's burial gown and distributed the small scraps of cloth to the officers, including Arnold.

Just a few months earlier, Arnold's army had successfully captured the British fort at Ticonderoga, New York, but now a much more daunting challenge faced them. They planned to march through Maine in winter and attack the city of Quebec, and the officers hoped that the small pieces of fabric would serve as talismans in the coming campaign.

Whitefield, a staunch man of God, never would have expected that one day he himself would be seen as a sort of patron saint of an armed insurrection. Thirty years before the American Revolution, he and two other prominent pastors, Jonathan Edwards and Gilbert Tennent, had transformed American life with their sermons. They had hoped to change the way Americans thought about their relationship with God, and they had succeeded, leading to a great religious revival that became known as the Great Awakening. But the religious movement had done much more: A generation later, it would provide many of the philosophical underpinnings of American democracy.

❧

Like many popular movements, the Great Awakening didn't really have a defined beginning or a definite end. The religious revival first appeared sometime around 1720, and after a few years it seemed to just fade away. Those who participated in the movement believed that every man is the same in the eyes of God—that is, every man is a failed sinner. The followers were known as New Lights, and they believed that everyone should strive to lead moral, useful lives and that education was necessary to serve and understand God.

The best known of the New Lights was Jonathan Edwards, a remarkable man who had gone to Yale when he was thirteen years old and by the age of twenty-one had become the head tutor there. Edwards is familiar to students of early American literature for his famous sermon "Sinners in the Hands of an Angry God" and for its well-known opening,

"The God that holds you over the pit of hell, much as one holds a spider or some loathsome insect over the fire, abhors you, and is dreadfully provoked."

When he would launch into one of his fiery messages, Edwards didn't make any effort to temper his harsh words with soothing glances; he preached while staring above the audience, never taking his eyes off of the bell rope that hung from the steeple at the back of the church. Even as the people rushed to the pulpit and cried for Edwards to stop, he continued preaching his damning message. Instead of rejecting Edwards's uncomfortable words, thousands of believers were drawn to the message in waves of people that astounded nearly all who witnessed the phenomenon.

Edwards, whose rebellious Puritan Congregationalists would take the name "Baptists," wasn't the only person preaching a new way of looking at life and at the church that dominated colonial society. Across the Atlantic in England, John Wesley, leader of a new group of believers, was preaching a new type of personal involvement in spiritual life. Because Wesley's group practiced a specified method of scheduling spiritual activity, such as fasting on certain days of the week, they soon became known as Methodists. George Whitefield was a colleague of Wesley's, and he traveled to America thirteen times to present this new form of Christianity. Whitefield was a skillful preacher—some claimed that Whitefield could make people swoon just by saying "Mesopotamia"—and he became America's first widely known celebrity.

The reaction of one Connecticut colonist, a man named Nathan Cole, was typical of that which greeted Whitefield when he arrived in an area. Cole recorded in his journal, "I was in my field at work. I dropped my tool that I had in my hand and ran home to my wife, telling her to make ready quickly to go and hear Mr. Whitefield preach at Middletown, then ran to my pasture for my horse with all my might, fearing that I should be too late."

Of course, despite its terrific popularity, the Great Awakening movement had its doubters. Benjamin Franklin attended one of Whitefield's sermons to see if what he had been hearing was true. Before he went, however, he vowed to himself to drop nothing in the collection plate. "I had in my pocket a handful of copper money, three or four silver dollars, and five pistoles of gold," Franklin wrote in his autobiography. "As he proceeded I began to soften and concluded to give the coppers. Another stroke of his oratory made me ashamed of that and determined me to give the silver; and he finished so admirably that I emptied my pocket wholly into the collector's dish, gold and all."

Franklin also doubted the size of Whitefield's audiences, which the believers boasted were up to twenty-five thousand people. Franklin didn't believe that a man could be heard above a crowd of that size, certainly not to the outer edges, and in his usual scientific way he set out to investigate. Traveling to hear Whitefield preach from the steps of the Philadelphia courthouse, Franklin walked to the edge of the crowd and began his calculations: "Imagining then a semicircle, of which my distance [from Whitefield] should be the radius, and that it was filled with auditors, to each of whom I allowed two square feet, I computed that he might well be heard by more than thirty thousand." In his newspaper, Franklin noted that the effect of Whitefield's preaching was so persuasive that "one could not walk thro' the town in an evening without hearing psalms sung in different families of every street" replacing "idle songs and ballads" throughout the city.

Whitefield's ability as a preacher was such that one German immigrant said that after listening to one of his sermons, she had never felt so inspired. Remarkably, the woman didn't understand English. A New York newspaper reported the effect that this spiritual revival was having on colonial life: "We hear from Philadelphia that since Mr. Whitefield's Preaching there, the Dancing-School and Concert-Room have been shut up as inconsistent with the Doctrines of the Gospel."

The Great Awakening reached across the entire spectrum of colonial society. Jonathan Edwards noted that even the elite joined his crusade: "Some that are wealthy, and of a fashionable, gay education; some great beaus and fine ladies" were included. But as remarkable as it was for the well-off to humble themselves, even more surprising were the other people that the followers brought into the movement. Women were invited to be leaders in services and in the movement, and Native Americans and African Americans were welcome to participate fully as well.

The New Light movement went much farther than simply inviting social undesirables into the services, though. Gilbert Tennent began preaching that many of the Presbyterian ministers were themselves destined for hell, undermining not only the religious order but also the social order and authority in many of the colonies.

Understandably, many religious leaders and lawmakers were in a panic. In Boston, the Harvard faculty members, who had close ties to the established church, were horrified at this common movement, calling it a "crisis." Connecticut outlawed the itinerant New Light preachers from preaching or performing marriage ceremonies. These measures did little good: When Elisha Paine, a popular New Light preacher, was jailed, he continued giving sermons from his cell. The crowds that came to listen grew so large that bleachers had to be built outside to contain them all. Still the established church leaders in Connecticut persisted—in one county, so many New Light followers were arrested for not paying their tithes to a proper church that a second story had to be added to the local jail.

In Virginia, a New Light Baptist preacher, remembered only as Brother Waller, had his sermon interrupted by the parson of the local church and the sheriff. Waller continued speaking and the pastor and the sheriff began horsewhipping the preacher's face and hymnbook. When Waller began to lead his outdoor congregation in prayer, he was jerked from the stage, his head was beaten against the ground, and the sheriff carried him to a nearby gate, tied him to it, and gave the preacher twenty lashes with his whip. When the beating was over, Waller returned to the stage and began preaching again.

In several cities the local authorities interrupted Baptist meetings, and the preachers were hauled off to jail. One preacher described how he had been charged with "carrying out a mutiny against the authority of the land." The Baptists responded that "We concern not ourselves with the government . . . we form no intrigues . . . nor make any attempts to alter the constitution of the kingdom to which we as men belong."

Nonetheless, by taking on the officially sanctioned church and criticizing its foundations, Jonathan Edwards, George Whitefield, and Gilbert Tennent *did* take on the government. Theirs was a Christianity both diverse and creative, not limited or checked by the government, not beholden to Rome, the local magistrate, the mayor, the King of England, or any other European power; a faith that relied on equality, voluntary participation, and lay leadership. This new belief held that no man was any more important in the eyes of God than Edwards's allegorical spider suspended over the fire—and that all were equal in their insignificance. Kings were not divinely chosen. The aristocracy was not, by nature, blessed. The rich were not fated to be rich and the poor were not fated to be poor. If all men were equal in the eyes of God, it made the government-imposed restrictions imposed on the worshippers difficult to accept.

WHAT IF?

But what if Whitefield, Edwards, and Tennent had been cowed by the opposition and decided to abandon their radical message and beliefs?

In his church in Northhampton, Massachusetts, Edwards felt moved by the spirit of God to preach a new message. He entered the pulpit, bracing himself with out-stretched arms. He began preaching, raising his voice, telling the congregation that every person, man or woman, white or black, king or commoner, is a sinner and is therefore as despised in the eyes of God as an arachnid is in the eyes of a man.

Three women from the congregation suddenly rushed toward the altar and begged Edwards to stop. This had not been the first time he had provoked such a response. Edwards knew that he was disliked in the community. While in other pulpits he'd been attacked and physically threatened. One man even brought a noose and waved it from his pew. He knew that other pastors in the area had been conspiring to stop him from preaching his message. But Edwards felt strongly about the righteous-ness of his message and had always pushed on, deter-mined.

This time, however, to his own surprise, he stopped. He looked out over the people in the pews and saw them as if for the first time. Many were weeping, both women and men. He saw the guilt and anguish in the eyes of his parishioners. "Judge not," Christ's words from Matthew came to him, "lest ye be judged." Right then, Edwards decided he could no longer cause people hurt for the ways that they had been living their lives. "This church was created by the divine hand of God," Edwards thought to himself. "I have no right to destroy what He has created. And if I proceed I am no better than the Serpent of Eden." Without finishing his sermon, Edwards bowed his head and walked slowly away from the pulpit and down the steps to the congregation.

Edwards didn't soften his views, nor did he leave the church. But he never entered the pulpit again. On nearly every Sunday morning he could be found coiled in a middle pew, in his usual spot on the outside seat. During the liturgy he mouthed the words of peace, but his heart was full of bile and bitter self-pity. The church had spurned his vision of how God's people should live.

Meanwhile, in England, George Whitefield had followed John Wesley's advice to try to spread the new message of Methodism in the colonies. The people in America warmly received Whitefield, but Whitefield had a flaw that destroyed his mission. He wanted to be loved by all. No, more than that: he wanted to be adored—adored by throngs of people.

In Philadelphia, the local clergy began preaching that Whitefield was a false prophet, someone not to be trusted. Whitefield was shaken. He had been ridiculed in England, but he didn't expect a harsher reaction in America. He expected the sinners in the colonies to object to his message, but why were the other men of God casting stones at him? What had he done to deserve such condemnation? After speaking to an unheard-of crowd of thirty thousand people, Whitefield met with a person who he knew had been writing pamphlets opposed to him and his message, a local editor named Benjamin Franklin. Franklin surprised Whitefield by explaining that although he had attended the sermon expecting to disprove the popularity of the movement, "I emptied my pockets, copper, silver, and gold, and I have computed that there are at least as many believers there as were at our Savior's sermon of the loaves and fishes."

Whitefield didn't hear a word of what Franklin was telling him. He was still thinking about the insults he had received from the day he first entered Philadelphia. He decided that it was time for him to shake the dust of the colonies from his shoes and return to England. "You Colonists, I suspect, are not yet ready to receive the Word of God," Whitefield declared to Franklin. "In America, the hearers of the Word concern themselves more with the afternoon shooting of turkeys than in considering their place in the hereafter."

"Ah, I agree," replied Franklin. "A fine, noble bird, the turkey. A bounty of blessings are manifest in that one bird. . . ." As Franklin continued his monologue on the merits of the turkey, Whitefield decided that there was no hope for the Americans and realized that he must return to England and stay there.

The "Great Awakening" was ignored by most of the colonies, so it was never given that sobriquet. Life in the British colonies in America continued just as it had for decades, until the mid-1770s.

After a series of insults against the business and freedoms of the colonists by King George III, some members of the Continental Congress decided to proclaim independence from the king and from England. No colony in all of the world had ever broken away from its native land,

and no people had ever proclaimed their equality to the king (at least not for long). A document saying just these unthinkable words was written by a Virginian, Thomas Jefferson, and offered to the members of the Congress for unanimous approval.

Although nearly everyone in the Congress and in the colonies was exasperated at the decisions of the king, not all of the members of the Continental Congress agreed with the document. The first sentence of the second paragraph, in fact, seemed to get to the crux of the problem. "We hold these truths to be self-evident," the sentence read, "that all men are created equal, that they are endowed by their Creator with certain inalienable Rights, that among these are Life, Liberty and the pursuit of Happiness."

"This has the distinct odor of blasphemy!" one of the delegates to the Congress shouted in the debate. Several delegates pointed out that the Scriptures didn't say anything about men being equal—quite the opposite, it appeared to assume that some men would be masters, some men followers, some men slaves. One of the delegates walked to the Bible that had been placed at the front of the room on the first day and began reading from the book of Colossians. " 'Slaves, in all things obey those who are your masters on earth, not with external service, as those who merely please men, but with the sincerity of heart, fearing the Lord.' There, gentlemen, there," the delegate said. "Writing that all men are created equal when common sense tells each of us that it simply isn't so. Rebelling against the divine right of our king. Ignoring the pleadings of our fine and learned pastors. It is all too much, gentlemen. I say, too much!"

Colony after colony voted to approve the document;

New Jersey and New York held firm for the king. In New Jersey, one-third of the colonists were soundly opposed to independence, and most of the remaining two-thirds were undecided on the subject. It was too little support for the delegates to the Continental Congress to vote for Jefferson's document, and the delegation from New Jersey voiced its opposition to independence. Emboldened by New Jersey's stance, the delegates from New York likewise voted against the declaration of independence.

The businessmen of New England and the slave owners of the South knew that if the colonies were divided by a British stronghold in New York and New Jersey, they could not carry out a war of independence. As quickly as the idea of independence had roared through the Congress just weeks before, a new slap of reality brought calls for a document expressing continued loyalty to King George. On July 4, 1776, the members of the Continental Congress signed the famous "Declaration of Continuity" that reaffirmed the colonies' allegiance to the king.

It's obvious that the battle over the Declaration of Independence didn't turn out that way; it's equally obvious that the revolution was not easily won. Not every New Light believer picked up the cause of the revolution. Even Arnold's march on Quebec ended in failure, despite his homage to the Great Awakening. Some said later that it was doomed because of the un-Protestant-like superstition of taking talismans into battle. Others thought that it hadn't worked because the ceremony had been undemocratic: Only officers were given pieces of Whitefield's burial clothing.

But an enormous change had taken place among the people of the new nation, and with fortunate timing. In June of 1776 New Lights in Pennsylvania replaced the pro-British Anglican leaders who had dominated politics there; similarly, New Lights took a leading role in politics in New Jersey. The result was that as New Jersey was wavering on the Declaration of Independence, a change in the legislature brought Princeton president John Witherspoon and a new group of delegates to the Continental Congress in Philadelphia in July of 1776, just in time to vote "yea."

The leaders of the independence movement certainly weren't the only ones affected by the Great Awakening, which had taken place a generation earlier. It was said that whole congregations of believers followed their preachers into battle, the men certain that theirs was a righteous cause. The Revolutionary War wasn't a clean, simple war. Next to the protracted Vietnam War, it was the longest war in America's history, stretching out for more than eight years. The people of the colonies were willing to fight and die, to send their sons and husbands away never to return because the people believed that they were fighting for more than their own comfort and wealth.

A common scripture in New Light meetings was Revelations 21:5: "Behold I make all things new," a scripture that could have been applied to their country as well as to their personal circumstances. The great patriot John Adams said, "The Revolution was effected before the War commenced. The Revolution was in the mind and hearts of the people; and change in their religious sentiments of their duties and obligations."

Chapter 2

Washington Retreats to Philadelphia

❧

Humiliated after losing New York to the British, the Americans were forced to retreat. What if George Washington and his troops had not crossed the Delaware River to fight the British again?

THESE are the times that try men's souls . . ."

In the cold December air a Continental soldier read aloud for the first time the now-famous words that Thomas Paine had written just days before. "The summer soldier and the sunshine patriot will, in this crisis, shrink from the services of his country; but he that stands it now, deserves the love and thanks of man and woman."

The men listening knew all too well that these were difficult times. Too many of the American soldiers standing there were wearing rags, too many of them were standing in the bitter air in bare feet, and too many of them didn't have any ammunition. It has been said that words do not feed starving men, but the words of Thomas Paine seemed to lift the spirits of this beleaguered corps. Perhaps it was because Paine himself had been with Washington's army during recent defeats at the hands of the British; perhaps because Paine was able to eloquently express what the soldiers themselves believed. "Tyranny, like Hell, is not easily conquered," Paine wrote. "Yet we have this consolation with us, that the harder the conflict the more glorious the triumph. What we obtain too cheap, we esteem too lightly. 'Tis dearness only that gives everything its value. Heaven knows how to set a proper price upon its goods; and it would be strange indeed, if so celestial an article as FREEDOM should not be highly rated."

Later, in his makeshift quarters, Washington was able to overcome his doubts about himself and his army. He decided that the moment to save the fledgling revolution was at hand. As Washington met with a visitor, Congressman Benjamin Rush, the general was distracted and kept writing words on bits of paper. During the visit one of the pieces of paper had fallen to the floor, and one of Washington's officers had retrieved it for Rush. On it Washington had written the words that would be the password for the coming days: "Victory or Death."

As the Revolutionary War had begun in the summer of 1775, George Washington had been named commander in chief of the army of the colonies. There was some concern about whether the forty-four-year-old Washington was up to the task. He had little military experience, certainly no experience moving entire armies around on a battlefield, but he was known for being bold and a stern disciplinarian. When he accepted the command, Washington instructed "every gentleman in the room" to remember that he himself had said that he was not fit for the honor.

Through 1775 and into 1776, Washington had delighted his supporters with victories in several battles, including the capture of Boston from British forces. Washington's successes no doubt emboldened the Continental Congress to declare American independence from Britain in July 1776, but almost immediately British resolve seemed to strengthen. A blunder by Washington on Long Island, north of New York City, caused a loss of five thousand soldiers. Washington avoided losing his remaining forces by escaping under the cover of a heavy fog into Manhattan. But losing two battles in New York—at a cost of an additional three thousand men—forced Washington to lead his army in a retreat. The redcoats chased the American army the entire length of New Jersey, and the Americans were able to save themselves from complete surrender only by crossing the Delaware River into Pennsylvania and taking every available boat with them to prevent the British from following.

With New York and New Jersey now under the British flag and the American army in disarray, Congress, on December 11, called for a national day of fasting and prayer, resolving that "all the members of the United States, and particularly the officers civil and military under them, [begin] the exercise of repentance and reformation." To demonstrate their faith in Washington in the face of a possible British attack on their location in Philadelphia, Congress passed another resolution: "Whereas a false and malicious report hath been spread by the enemies of America that the Congress was about to disperse: Resolved, that General Washington be desired to contradict the said scandalous report in general orders, this Congress having a better opinion of the spirit and vigour of the army, and of the good people of these states than to suppose it can be necessary to disperse. Nor will they adjourn from the city of Philadelphia in the present state of affairs, unless the last necessity shall direct it."

The very next day the fine men of Congress changed their

minds and fled Philadelphia for the safer confines of Baltimore. Once they had settled in there, they instructed Washington to strike from any written record of his orders their resolution of the previous day.

Congress wasn't the only group to demonstrate less than full confidence in Washington's efforts. Well-known patriots were disavowing the new government and pledging allegiance to the king. The New Jersey legislature disbanded so that each legislator could try to save his own neck, and across New Jersey people were lining up to receive royal pardons for their rebellion by swearing allegiance to the king. When Washington's troops went to local farmers to buy food, the farmers turned them away because the only thing the soldiers offered in return was Continental money, which the farmers felt was about to become worthless paper.

Washington himself was no fountain of optimism, writing, "I think the game is pretty near up, owing, in a great measure, to the insidious Arts of the Enemy, and disaffection of the Colonies before mentioned." A few days later he wrote in another letter, "We find, Sir, that the Enemy are daily gathering strength from the disaffected; this Strength like a Snow ball by rolling, will Increase, unless some means can be devised to check, effectually, the progress of the Enemy's Arms . . ."

With Washington safely under control, British general William Howe decided to return to New York and left the fort at Trenton, New Jersey, under the command of Colonel Johann Rall. Rall was not much of a soldier, although he loved the accoutrements of the military. He never asked the men about the conditions of their weapons or about preparing fortifications but instead busied himself with the condition of the regiment's band, which was his pet love.

Rall's actions weren't completely negligent considering the information that was provided to him. One of the British generals [wrote] to him: "You may be assured that the rebel army in Pennsylvania . . . does not exceed eight thousand men who have neither shoes nor stockings, are in fact almost naked, dying of cold, without blankets, and very ill supplied with provisions."

Even after Rall was warned by spies that the Americans were considering an attack, he dismissed the reports as "old women's talk." "These clod-hoppers will not attack us, and should they do so, we will simply fall on them and rout them," Rall boasted.

Washington decided not to wait for spring to counterattack the British, as might have been expected, but instead chose to attack the fort at Trenton on Christmas Day in 1776. Washington calculated that the British would be celebrating the holiday instead of watching their posts and that over half his own army would most likely begin walking home in less than a week because their obligations to Washington's army would be complete.

To reach the fort at Trenton, the American forces had to cross the half-mile-wide Delaware River, which was full of ice floes that could break up the small boats they would use. To make the situation worse, on the chosen night the weather seemed to be against them. A Lt. Col. John Fitzgerald recorded in his journal, "It is fearfully cold and raw and a snow storm setting in. The wind is northeast and beats in the faces of the men. It will be a terrible night for the soldiers with no shoes." Another soldier reported, "Troops began to cross about sunset, but the force of the current, the sharpness of the frost, and a high wind rendered the passage of the river extremely difficult."

Washington had wisely assigned the task of getting the boats across the river to a Massachusetts regiment called the Marbleheaders: experienced men—black as well as white—from whaling ships. The regiment was led by Col. Henry Knox, a tall, burly man who in civilian life had been a struggling bookseller. Knox and his able crew were able to ferry Washington's troops across the eight hundred yards of ice-choked water without losing a single boat, but even more important, were also able to transport eighteen pieces of artillery across the river.

Once the men were safely across the river, they faced a cold nine-mile march to Trenton. Two soldiers in fact died of exposure on the march through the snowstorm, and many more were worn down by the effort. A young fifer said that at one point he became so exhausted that he sat down on the stump of a tree and began to pass out; his sergeant found him and made him walk around until his body was warm again. But the snow that had made their crossing so treacherous and their march so difficult was in fact a blessing. The storm had muffled the sounds of the army moving toward the town, and as they encountered the British troops, the snow was blowing from the south and into the faces of the enemy.

The fighting started on the edge of Trenton. Soldiers on both sides found that the snowstorm had wetted their gunpowder, rendering their guns almost useless. The Americans fixed bayonets on their guns and rushed toward the Hessian soldiers. The Hessians managed to get off a few sparse shots to no effect, and when they found that most of their guns wouldn't fire, they turned and ran back toward town like "frightened devils." The streets of the city were not the refuge the Hessians sought. The American soldiers had moved into the city and begun firing their artillery into the narrow streets. (Knox and his men had wisely covered their artillery so that these weapons were dry.) The civilians of Trenton, who had recently pledged loyalty to the British, now turned on them and began firing at the Hessians from the buildings. As the well-drilled Hessian soldiers tried to form ranks, they became easy targets for the cannon in the narrow streets. In all, more than one hundred Hessian soldiers were killed and more than nine hundred were taken prisoner.

Colonel Rall, who had dismissed the courage of the Americans, was slow to react to the attack. Even after a Hessian sentry ran to his headquarters and shouted *Der Feind! Der Feind! Heraus! Heraus!* ("The enemy! The enemy! Turn out! Turn out!"), Rall did not respond. The sentry returned to find Rall still in his evening clothes, and when the sentry returned a third time, he found Rall just then in the process of getting dressed. "What is the matter?" Rall asked, still not understanding that his fort was under full attack. A favorite old American story, perhaps apocryphal, claims that Colonel Rall had made an unfortunate decision earlier that evening. While in the midst of drinking and enjoying his card game, he had been handed a note from a sentry that reported that the Americans were attacking with their full force. Instead of interrupting his card game to read the note, Rall shoved it into his pocket and continued to enjoy his evening. According to the tale, he finally read the note as he lay dying and realized that if he had stopped playing cards long enough to hear what the sentry had to say, he could have stopped the American offensive.

True or not, Rall's response to the attack could not have been worse. A Hessian captain reported later that "to his good fortune, Colonel Rall died the same day from his wounds. I say this because he would have lost his head if he had lived."

⤔

The Americans had won an important battle at Trenton, but Washington still faced the possibility of losing his army. Many of the men still were due to be discharged on December 31, and Washington knew that British general Howe would not let the insult at Trenton stand without a counterattack.

On the 29th of December, Washington took his men back to Trenton, to the place of their victory. Offering the men a bonus of ten dollars if they would stay an extra six weeks, Washington addressed them from his saddle and asked for volunteers to stay. He then rode to the side of the troops and ordered the drummers to play as he waited for men to step forward. Not a single soldier volunteered. Washington then rode back in front of his men and spoke to them again.

"My brave fellows, you have done all I asked you to do, and more than could be reasonably expected, but your country is at stake: your wives, your houses, and all that you hold dear," he said. "You have worn yourselves out with fatigues and hardships, but we know not how to spare you. If you will consent to stay you will render that service to the cause of liberty and to your country which you probably never can do under any other circumstances. The present is emphatically the crisis which is to decide our destiny."

Washington rode to the side of the troops again, and the drummers again picked up their rat-a-tat-tat. A few individuals stepped forward, and then small groups until finally nearly every man who thought he was fit enough to serve—more than a thousand in all—had volunteered to stay. For these men it was truly a brave act. No doubt many of the men knew the risks of what they were doing, and the risk was grave: Within a few weeks more than half of the volunteers would be dead from battle wounds or disease.

Late in the day on January 2, Gen. Lord Charles Cornwallis arrived in Trenton with eight thousand British soldiers eager to seek revenge. There was a short skirmish, but the Americans were able to pull back across Assunpink Creek, and although the British feigned following the Americans, darkness was falling and they also pulled back to their camp to prepare for the next day's battle. Cornwallis was in no hurry to attack the Americans, who were caught between the creek and the Delaware River. "We've got the old fox safe now," Cornwallis said. "We'll go over [the creek] and bag him in the morning."

It wasn't an overconfident boast. The British had eight thousand men to take on the fifty-five hundred or so American soldiers, they had better equipment, they were rested, and they had a superior position. Washington then made another bold move.

That night Washington ordered fires built for the camps and for the camp to be set up as noisily as possible. Instead of allowing the men to bed down for the night, however, at midnight Washington ordered them to march. The men were confused: Not only had they been denied sleep, but they weren't told where they were marching to or whether, like in New York, this was another retreat. Washington had the men cover the wheels of the artillery and the hooves of the horses in cloth so that they would not be heard as they traveled. Meanwhile, he ordered that the campfire be kept burning all night and that a force of men begin building fortifications across from Assunpink Creek, where the British would cross in the morning.

Washington's ruse worked. When Cornwallis awoke the next morning the American forces were gone. He would have howled even louder had he known where Washington was marching—back to Cornwallis's supply base in Princeton.

The Americans expected the supply base to be deserted, but to their surprise, when they were about a mile from Princeton, they were met by a large force of British soldiers. Washington ordered his men into battle, but they moved toward the enemy reluctantly and then broke ranks and retreated from the advancing redcoats. It appeared that, as in New York, they were about to be routed again. Suddenly Washington himself arrived, riding his white horse, and called out, "My brave fellows, there is but a handful of the enemy, and we will have them directly!" With that, he rode to the front of the line of American troops and waved for his men to follow him. The general rode to within thirty yards of the British, and while in front of his troops, and himself still between the two lines of muskets, Washington ordered his men to fire.

Washington's aide covered his face with his hat so that he would not have to witness the death of the general. When the guns quieted, the aide looked to see Washington still on his white horse and the British running from the fight. Washington told his aide, "Away my dear Colonel, and bring up the troops. The day is our own."

Washington had hoped to capture all of the supplies at Princeton, but when word was delivered to him that Cornwallis was on his way from Princeton, "running, puffing, and blowing and swearing at being so outwitted," Washington took the time only to find some better blankets for his men before marching deeper into New Jersey. Cornwallis, stung by the losses and confused about where Washington might appear next, did not pursue the Americans any further.

Frederick II, King of Prussia, known more popularly as Frederick the Great, called Washington's maneuvers in New Jersey "the most brilliant of any recorded in the annals of military achievement." More important than the strategic value of the victories, however, was the effect the reversals of fortune had on the British army. "A belief prevails," Washington wrote, "that the enemy are afraid of us."

WHAT IF?

What if George Washington had decided not to bet his nation, his army, and his own life, placing everything in balance between victory and death?

On Christmas night, 1776, George Washington sat alone in his tent wondering how long the storm would last, wondering if by some miracle he could hold his army together, wondering if he would ever see his plantation at Mount Vernon again.

Although he had asked his friend Thomas Paine to write an inspiring message for the troops, hoping that the men would have their spirits boosted and that more than a few of them would decide to reenlist, Washington knew that the men needed more. The men needed blankets, shoes, and food, and most important, hope, a well-founded hope that they could overcome. The men doubted themselves, the local citizens doubted, and in the worst blow of all, even the leaders in Congress doubted that they could hold the British back.

It was critical that Washington protect Philadelphia, even with the Congress taking up residence in Baltimore. If the British were to cross the Delaware River, they would have a quick thirty-mile march to Philadelphia. Philadelphia was the largest city in the nation at that time, and because Congress had been meeting there, it was also the seat of the fledgling government. No city was as critical to the success of the United States, and the fall of the city would most certainly mean the beginning of the inevitable end.

Under the whistling notes of the wind of the winter storm that had just begun, Washington could hear the distant bass of the cannon of the Hessian soldiers. Yet another celebration, this one bigger than most because of the holiday. No doubt they were happy, laughing, warm. His own men . . . his own men had given more than could be expected, and more yet was needed. He couldn't wait any longer to see what the British intended to do. He had to make a decision now.

The night after Christmas, Washington ordered his men awakened at midnight. A few were told to keep the campfires burning through the night, and the rest were ordered to form ranks to march. They were going to Philadelphia.

Another retreat wasn't Washington's first choice, but he had no other practical option. Half of his army would be going home in less than two weeks as their term of service expired, and he had no hope of holding back the British army. If he could get to Philadelphia, he could find a strategic position, erect fortifications, and have a chance against the redcoats when the battle of the spring arrived.

The long march was an ordeal of cold and hardship. A winter storm had arrived the night of the march, and by dawn eight men had died from exposure. Through the next day the men endured a difficult march over snow-and-ice-covered roads that turned to mud during the daylight. His men were exhausted, and to make matters worse, the townspeople were either indifferent, with many farmers refusing to sell food to the soldiers, or outright hostile toward the army because of the hardships the war was bringing to the citizens. The soldiers had to endure catcalls and shouts from several farmwives along the way.

The army finally arrived at the outskirts of Philadelphia and began making a new camp. As soon as he had settled into his tent, Washington received a message informing him that "the necessary removal of the army to Philadelphia has rendered the situation in Baltimore of concern, and therefore the Congress will be convening in Richmond." The note went on to express that several members of Congress planned to stop by Washington's plantation at Mount Vernon and pay regards to Martha Washington, a gracious gesture that nevertheless brought an air of sadness upon the general. "Although I have no doubts of our Cause, I do doubt that we have the Means to execute the necessary actions required," Washington wrote. "The support from Congress is not adequate to sustain our army, and no other source of sustenance appears likely."

The Revolutionary army was able to collect blankets, shoes, and warm meals in Philadelphia, and they soon began the work of building fortifications necessary to defend the city against the British. The spirits of the now-warm and well-fed American troops were greatly improved until surprising news arrived from Baltimore.

British general Lord Cornwallis had been furious with Hessian colonel Rall for letting Washington slip away on Christmas night, and he decided that he would not pass the harsh weather in winter quarters, as was the European custom, but would march his forces from Princeton into Maryland. The premonitions of Congress had been correct. Cornwallis's army soon overran the unprepared city of Baltimore.

From their new site in Richmond, Virginia, the members of Congress decided that the distance from Baltimore to Richmond didn't provide the cushion that they desired, and they informed Washington that for the good of the nation, they were planning to reconvene in Charleston, South Carolina. Washington had more to worry about than the safety of Congress, however. The Hessian troops were now off to the northeast in Trenton, and Cornwallis and his army were to the southeast in Baltimore. The two armies could act as pincers, attacking from two sides. Washington couldn't hope to hold on against two armies, both larger than his, attacking at the same time. Before the fortifications that were to surround Philadelphia were complete, Washington ordered his army to march to the west, into the mountains of Pennsylvania. From there he could hope to hold his army together through the winter to attack again in the spring.

Concerned that Washington had left Philadelphia to the British, the members of Congress decided to leave Charleston for the warmer—and, they hoped, safer—city of Savannah, Georgia. Congress wasn't the only group of citizens dismayed by the events of the winter of 1776 and 1777. All across the colonies, citizens were abandoning the cause of the revolution, and even in Washington's mountain hideaway, soldiers simply walked away from camp in groups of a dozen or more. By the time the warm air of spring blew and Colonel Rall was finally ready to leave winter quarters to join in the pursuit of Washington, the rebellion was over.

Washington was so thrilled with the victory at Princeton that he thought he could end the war that very day. It was not to be, and the Americans—and Washington—were to suffer many more horrible battles and losses before realizing victory seven wrenching years later. Late the next summer, Philadelphia did fall to the British, but they were forced to abandon the city a year later, and the British army returned to New York after fighting for two years and gaining nothing. That was how the Revolutionary War went, the British fighting, winning, and then being forced to pull back. After seven years of this, they had had enough and decided to give the Americans their freedom.

To hold out against the British, the Americans had to have a resolve that they could win, and that resolve was gained when Washington crossed the Delaware River to face the British at Trenton. An English journalist reported the effect of the victory: "The minds of the people are much altered. A few days ago they had given up the cause for lost. Their late successes have turned the scale and now they are all liberty mad again. . . . [N]ow the men are coming in by companies." Emanuel Leutze immortalized Washington's victory eighty years later in his famous painting of the event (although Leutze made several errors, including painting Washington as a sixty-five-year-old president, instead of the forty-four-year-old general).

By standing his ground in New Jersey, Washington showed the British that the American Revolution wasn't over, as they had thought, but was just beginning.

Chapter 3

The Loss of Fort Sackville

∽∾

In a battle that was little noticed at the time, and is hardly remembered today, George Rogers Clark defeated the British on the western frontier and captured Fort Sackville at the small village of Vincennes. What if Clark had failed?

PEOPLE living in the thirteen colonies in the 1770s were furious with the British over the price of tea, local tax issues, and other sundry insults that caused them to request, at the end of a musket, that the British kindly take a smaller participatory role in government. But such complaints must have seemed petty to the American pioneers living in the Kentucky and Indiana territories, a region known at the time as the Northwest frontier. These colonists had a far worse grievance.

Since the mid-1770s, the British had organized a war of terror against Virginians who were crossing the Appalachian Mountains and settling in what is now Kentucky, Indiana, Illinois, and Ohio. The most hated of all the British was Henry Hamilton, the lieutenant governor of Canada, whom the Americans called the Hair Buyer because he paid rewards for each scalp of a Virginian settler that the Native Americans delivered to him. Hamilton's commanders in England had even shipped eighteen hundred steel knives to Hamilton at Fort DeTroit to distribute to the Indians to make the scalping easier. In one month in 1777, the Native Americans delivered 129 scalps to the Hair Buyer, and for generations in Kentucky, 1777 was known as the "dark and bloody year."

A twenty-six-year-old Virginian, George Rogers Clark, decided that he could do something about the situation on the frontier, and in 1778 he convinced the fledgling United States to help fund an army to go to the Northwest frontier to confront the British and save the settlers there. Like fellow Virginian George Washington, Clark was a surveyor, and like another famous Virginian, Thomas Jefferson, Clark was tall, redheaded, and handsome. He was also the older brother of William Clark, who would later join Meriwether Lewis to explore North America, but George Rogers Clark would soon accomplish a feat that surpassed, in bravery at least, any of the better-known accomplishments of these other men.

Clark arrived in Kentucky at a camp near present-day Louisville in the summer of 1778. He carried orders signed by Patrick Henry, governor of Virginia, instructing him to raise a force of men to protect Kentucky from the British-supported Indian attacks upon the settlers there. But Clark carried a set of secret orders as well, also signed by Patrick Henry, that instructed him to launch a military campaign hundreds of miles north into British-controlled territory. Clark knew that the men might try to escape once the true nature of the second set of orders was revealed. "I was sensible of the impression it would have on many, to be taken near a thousand miles from the body of their country to attac a people five times their number, and merciless tribes of Indians their allies and determined enemies to us."

Clark intended to capture three important British forts:

two in the small settlements of Kaskaskia and Cahokia, which were located on the Mississippi River in what is now Kentucky. Then Clark planned to take Fort Sackville, located on the Wabash River in the French town of Vincennes, which is in present-day Indiana. With the advantages of stealth and surprise, Clark quickly took the forts at Kaskaskia and Cahokia, and when travelers told him that Fort Sackville was undefended by the British, he sent Capt. Leonard Helm and a platoon of men to Vincennes to claim the fort.

Arriving in Vincennes, Helm simply walked into the undefended fort, took down the British flag, wrapped it around a rock, and threw it in the nearby Wabash River. Clark had accomplished what he set out to do—capture all three of the British forts in the lower Mississippi Valley. This undertaking was the only successful American military campaign of the Revolutionary War in 1778, but these were hardly the kinds of victories that would cause the British to flee back to England. When word reached Henry Hamilton in Fort DeTroit (the forerunner of the city of Detroit) that Fort Sackville in Vincennes had fallen to the Americans, Hamilton quickly organized a force to retake the three forts.

Unfortunately, his campaign to retake Fort Sackville didn't begin as smoothly as he would have liked. Hamilton loaded his 140 men, their supplies, and a heavy piece of artillery (a large cannon capable of shooting six-pound cannonballs one hundred yards) onto a small barge for the crossing of Lake Erie. In their enthusiasm, the men fired the cannon as a parting salute to Fort DeTroit, but the cannon's recoil caused the barge to sink, and the campaign was delayed while men drove into the chilly water to retrieve the cannon and their supplies.

When word arrived in Fort Sackville that Hamilton was

on his way with a force of men, the local French militia that Helm had recruited abandoned the fort, leaving just Captain Helm and three other men to face the advancing British. Helm sent a letter to Clark (which never arrived because the messenger was captured) saying, "I am determined to act brave." As Hamilton's army approached Vincennes, Helm waited beside a loaded cannon, with a bottle of apple brandy in one hand and a slow-burning fuse in the other. Two townspeople convinced him not to fire the cannon because it would needlessly endanger their families over what was to be an inevitable loss, and Helm quietly surrendered to Hamilton.

Hamilton was dismissive of what he had won. "In this miserable picketted work called a fort was found scarcely anything for defense. . . . There were not even platforms for small arms or even a lock to the gate." Soon after their arrival, however, Hamilton and his men were able to turn the fort into something that resembled a military outpost. Hamilton had 500 soldiers and recruited another 250 marksmen from the local community to serve as a makeshift militia. Now all he had to do was to wait out the winter and be ready for the spring thaw, when his army could move again.

That winter of 1778–79 was warmer than many in the Midwest, and instead of snow, the skies opened up with rain, enough rain to flood the entire Wabash River basin for miles around Vincennes, forming a natural moat around the already well-prepared Fort Sackville.

In Kaskaskia, Clark learned of Hamilton's success at Vincennes and the large force that he had brought with him. Clark knew that Hamilton didn't need a large force of men and a piece of heavy artillery to recapture Fort Sackville—obviously Hamilton planned an attack on Clark and his army. Clark had two unpleasant choices: He could either march on well-defended Fort Sackville in the spring, or he could wait for Hamilton to recruit more Indians and attack at some unforeseen moment. Clark was outnumbered, and that situation was only going to get worse. He had fewer than 180 men, and he knew that the possibilities of reinforcements, or even communication, with the armies back east were impossible. "I would have bound myself seven years as a slave to have had five hundred men," he later said.

Clark decided on a third, almost absurd plan: He would march his small force across the 240 miles of flooded land in February and try to take Fort Sackville immediately. "I am resolved to take advantage of this present situation and risk the whole on a single battle," he wrote to the governor of Virginia. "I know the case is desperate, but Sir, we must either quit the country or attack Mr. Hamilton. . . . We have this consolation, that our cause is just and that our country will be grateful and not condemn our conduct in case we fail."

On Friday, February 5, the men set out on their 240-mile march. Two weeks into their march they encountered their first flooded plain, where the lowlands between the Little Wabash River and the Fox River had flooded waist deep. The nearest dry land was five miles away, and the next day they would have to march through February-cold water. But somehow the men remained in high spirits. That evening in camp, a drummer boy entertained them by floating on his drum, and the men began telling tall tales about the coming battle. "They wound themselves up to such a pitch that they soon took Ft. Vincennes, divided the spoil, and before bedtime was far advanced on their route to DeTroit," one of the officers recorded.

The mood would soon change. As the men marched through flooded fields, they watched many of their provisions float away or become destroyed in the water. The wet walk, combined with little to eat, had the men in deep gloom, made worse by the weather. It had begun to rain, and the temperatures, which had been quite moderate for February, had begun to fall. When they began their march, Clark had sent a flatboat, the *Willing*, up the Wabash River with supplies and provisions that they would need for the attack on the fort. Hoping to retrieve the provisions, Clark sent a party of men off on a hastily constructed raft to find the flatboat. Captain Bowman described in his journal that the search party was the group's last hope. "Starving," he wrote. "No provisions of any sort now for two days." The next morning the search party returned empty-handed—their raft had fallen apart, and they had survived by spending the night clinging to logs.

Still, four days after they had first begun crossing the flooded fields, Clark's army was just five miles from Fort Sackville, close enough that they could hear the daily cannon salute at the fort. But now Clark and his men faced the main flood plain of the Wabash River, where the water was almost certain to be deeper than they had already encountered. Clark hoped that the march wouldn't be as severe as it looked, and he set out in a canoe himself to test the depth of the water. He found that in many places the water was more than five feet deep.

When Clark returned to camp, he found his men desperate and hoping for some good word. "All ran to hear what was the report," he later wrote. "Every eye was fixed on me." Clark didn't dare tell his men just how much worse their situation was about to become. Instead, he whispered to some of his officers to do exactly what he was about to do. Clark scooped up a handful of gunpowder, and then poured some water into his palm. He then rubbed the makeshift warpaint on his face, gave a yell, and walked sharply into the water. His men followed him into the ice-cold water without saying a word.

Finally the men camped on an island, from which they could actually see the fort less than two miles away. That night cold weather set in and ice formed on the water's edge more than a half-inch thick, and the men still had nothing to eat. "No provisions yet," Bowman wrote. "Lord help us."

Clark's plan to attack the fort was no less bold than his march across the flooded fields in the middle of winter. "It was now that we had to display our abilities," Clark later wrote. "We knew that nothing but the most daring conduct would insure sucksess." First, Clark sent a letter into the town of Vincennes warning the inhabitants that the fort was about to be attacked by a force of one thousand men who had brought heavy artillery. He warned the townspeople to either stay in their homes, if they supported the Americans, or to seek refuge in the fort if they supported the British. Clark didn't worry about the letter being passed along to Hamilton, because of the lie about the size of the American force. Clark then marched his men to a point near the town, carrying enough different battle flags for a force six or seven times their size, with the flags mounted on extra-long eight-foot poles so that they could be seen at a great distance.

That first day there was no fighting, but the French settlers in Vincennes came out toward the army carrying plates of food. Both groups were grateful: For Clark's army it was the first food they had eaten in five days; as for the townspeople, they were so opposed to British rule that not one of them had joined Hamilton's men in the fort.

When darkness arrived, Clark's men began firing. The frontiersmen were such capable shots that the British soldiers were not able to use the gunports in the walls of the forts because the Americans were able to shoot through the holes. During the fighting Clark heard that a small British patrol had returned and that the men were hiding in the town. Clark ordered his men to pull back, and just as he had planned, the British soldiers crept up to the wall, gave the password, and began climbing up on ladders that were lowered from the fort for them. When the men from the patrol were partly up the ladder, Clark's men let out a battle cry, just to watch the soldiers jump over the wall in a panic. Clark knew that the patrol would pass along the rumor that he had planted in the town that the American force contained almost one thousand men.

Early the next morning Clark demanded that Hamilton surrender, but Hamilton refused. Then events turned gruesome. A patrol of Indians with particularly poor timing returned into the midst of Clark's army. Clark, wanting revenge upon the men for their raids on the settlers in Kentucky, and wanting to send a message to other Indian tribes that the British could no longer protect them, executed eight of the warriors in full view of the British soldiers in the fort.

The next morning Clark again pressed Hamilton to surrender, and this time, after having seen the kind of violence Clark was capable of, Hamilton agreed. When Clark and his tired men marched into Fort Sackville, Hamilton asked, "Colonel Clark, where is your army?" When the deception became obvious, Hamilton turned his back to Clark with tears in his eyes.

WHAT IF?

What if George Rogers Clark had not undertaken an absurd march across flooded fields in February to capture the British fort?

After taking Fort Kaskaskia, Clark received word that Henry Hamilton and a force of men had retaken Fort Sackville in Vincennes. The news was hardly surprising, although Clark was concerned about how Hamilton the Hair Buyer had treated his loyal Captain Helm. Clark had sketched out a plan to march as soon as winter arrived. "The hard freeze will allow the army to march quickly over solid ground," Clark recorded in his journal. Clark knew that the march would be difficult, but his soldiers were hardened frontiersmen who were used to living in the wilderness in the harsh Midwestern winters.

But as Clark waited, instead of winter cold the weather brought days and days of rain. Clark knew that the land would be covered with miles of nearly impassable mud, if not flooded with water. A march now would be foolhardy at best, fatal at worst. A march would surely decimate his small army even before he reached Fort Sackville. Clearly, he would have to wait until the water dropped in the spring before his army could march to Vincennes.

In April the water receded enough for Clark and his men to start out. It was a long, wet, difficult march, but they crossed the 240 miles of wilderness without incident. Or so they thought.

Hamilton's patrols of Native Americans were able to track the American force almost through their entire journey by the footprints they left in the muddy trails. Had they marched while the fields were flooded they would not have been discovered, because water leaves no footprints, but now the Indians could report almost daily to Hamilton the progress of the Americans and the size of their force.

Arriving in Vincennes, Clark tried a well-thought-out ruse, disguising the size of his army by using extra battle flags placed on extra-long poles. But Hamilton and his

British soldiers weren't fooled, because of the information that had been provided. As the Americans shouted and fired their guns outside the walls of the fort, Hamilton waited. Every few hours, when Clark's men were emboldened enough to make an assault on the walls of the fort, Hamilton would give the order to fire the six-pound cannon. Each time, the cannon fire would kill one or two of the Americans, sending the rest scattering back into the woods and giving Hamilton a good laugh. But Hamilton wasn't interested in engaging Clark's troops—he planned to leave that to the platoon of Indians who were waiting outside the city.

For two days Hamilton let the Americans run down their ammunition and enthusiasm by firing at the well-protected fort. Then, suddenly, a screaming, terrifying group of Native Americans came running from behind Clark's men, just as the gates of the fort swung open and Hamilton's men came running out. Within minutes Clark's small force was routed, and Clark was forced to surrender. The American excursion into the British Northwest was over.

Back in the colonies, most people were too busy to notice that Clark had failed out on the frontier. The men in Boston, New York, and Philadelphia had more pressing concerns to address. But in Kentucky and Tennessee, the news of Clark's defeat spread from settlement to settlement and cabin to cabin. The American settlers were near panic. They realized that they could stay and battle the British and patrols of tomahawk-wielding Indians alone, or they could leave. At first only a few families packed up and made the journey back through the Cumberland Gap in the Appalachian Mountains to Virginia. But during the summer of 1779, Henry Hamilton increased the rewards for scalps of Americans because he was angry at having spent the winter in the small village of Vincennes, and during the hot summer months more Indian raids took hundreds of American lives. By autumn, the trickle of settlers winding its way back through the Cumberland Gap had become a flood.

The loss of Fort Sackville had little effect on the Revolutionary War, and Washington's army was able to force Britain to recognize American independence in September of 1782. But like a small stone in a pond that causes ever-widening effects, Clark's failure spread through the history of America.

Because of Clark's inability to capture Fort Sackville, the western border of the newly created United States became the Appalachian Mountains. This natural barrier prevented violence between the Americans in the East and the British subjects living in the Canadian regions of the Ohio Valley. But a few low mountains couldn't contain the animus the two peoples felt toward one another, and eventually another war began.

The Second Revolutionary War, more commonly known as the War of 1812, was a serious affair. With loyal British subjects to the North and to the West, battles were frequent and bloody for the Americans. Having the greedy Spanish governors to the south and in the Southeast didn't give Americans in the Southern states much peace of mind, either. When British Adm. George Cockburn marched into the United States Capitol and ordered his men to put the torch, most Americans realized it was time to sue for peace at whatever the cost. The United States was forced to give up the state of Maine and portions of Vermont and New Hampshire. America survived, but it was a small country, one of many that eventually formed on the continent of North America.

Of course, it didn't happen that way. In the most heroic battle in a decidedly heroic war, Clark was able to defeat the British. Because of Clark's conquest of Fort Sackville, the raids on the American settlers in the Old Northwest ended. When the Revolutionary War was finally over, Britain granted the United States all lands east of the Mississippi River and north of the Ohio River in the Treaty of Paris. George Rogers Clark had added almost as much land to the United States as existed in the original thirteen colonies.

America Scraps
Its Constitution

❦

In the early years of the republic, the government of the United States was nothing but a "cobweb confederation." What if the government had continued under the Articles of Confederation, despite their many faults?

SOMEONE so desperate for entertainment that they begin thumbing through the appendix of a modern American history textbook will notice a queer thing: Although the appendix will no doubt have an impressive collection of documents and speeches that are important in the nation's history, there will be something missing. What these textbooks don't like to mention is the messy fact that our nation didn't proceed from one sacred text to the other, directly from the Declaration of Independence to the Constitution, forever and ever, amen. The missing document is called the Articles of Confederation, and it was the blueprint for our nation's failed, first government.

❦

In June of 1776 the Continental Congress decided to draft a document stating that the colonies were no longer subject to British rule. The document, written by Thomas Jefferson, was called "The Unanimous Declaration of the Thirteen United States of America" (the title "Declaration of Independence" wasn't used at the time) and was issued by the "thirteen united States of America"—little *united*, capital *states*.

Having said that the government of Britain no longer had authority in America, the men in the Continental Congress quickly began constructing a new government. The first attempt at a constitution, not surprisingly, had been by the great thinker Benjamin Franklin, who submitted a draft of a constitution a year before the Declaration of Independence was written. But it was John Dickinson, delegate from Pennsylvania, who wrote a constitution that was eventually adopted by the Continental Congress after a few revisions. In late 1777 the Articles of Confederation were submitted to the states for approval. There was some sniping over which states were going to get the land to the west, but that was eventually straightened out, and in 1781 the Articles of Confederation were finally ratified as the official constitution of the United States of America.

Men of mediocre talents didn't write the Articles of Confederation, and looking at it from their point of view, many of the elements make sense. The Articles of Confederation, with their emphasis on strong state governments and a weak national government, more closely followed the natural-law philosophy that had guided the Revolution than did the subsequent Constitution. It was an idealistic charter that depended on states' cooperation rather than federal coercion to get things done. There was also the pragmatic matter that the Articles were written during a war, and there was an urgency that some form of governance be quickly organized. No national government existed, but each colony did have some sort of governmental structure, and so giving power to the states was the most expeditious thing to do.

Under the bare bones of the Articles of Confederation, the individual states were able to provide for their own armies, enact their own embargoes, and even carry on individual negotiations with other nations. There was no executive branch or president, just a Congress that was given broad powers and in which delegates could serve no more than three years out of any six. The power in the national government rested not in individuals, such as a president, but in congressional committees. Under the Articles, Congress could not tax the citizens, was not allowed to regulate trade, and wasn't even allowed to pass laws that applied to individuals. The arrangement wasn't really a government at all, but more like a union of nations that could be brought together for the purpose of waging war, an eighteenth-century version of NATO.

Finally, the Articles required the approval of nine states to do anything significant and required the approval of all thirteen states to make changes in the Articles themselves. Author Noah Webster was dismissive of what the Articles had created: "So long as any individual state has power to defeat the measures of the other twelve, our pretended union is but a name, and our confederation, a cobweb." This provision of the Articles of Confederation almost brought about the end of the United States before it had even really begun.

By 1781 the American colonists had pulled the upset of the millennium, defeating the world's most powerful nation to gain their independence, and Britain agreed to give the troublesome Americans the right to govern themselves. The British had grown tired of sending men and money off to fight to maintain authority over a group of uncivilized people whom they didn't really care about, and a peace treaty was signed on September 3 and delivered to the Americans on November 30.

The problems arising from the weak Confederation began immediately. To ratify the treaty officially ending the Revolutionary War, nine of the thirteen states had to vote their approval within six months of its signing. But four months passed without nine states having representatives present in Congress at one time, and the treaty still didn't have the required signatures. Thomas Jefferson, chairman of the treaty committee, warned that there was a real chance that the treaty would not be signed by the deadline of March 3, 1782, and that the war could begin again.

Some people urged Jefferson to forge the signatures of the representatives of the missing states, suggesting that the British would never be the wiser, but Jefferson wasn't willing to found a nation on such a deception. One of the delegates was too ill to attend sessions of Congress, and as a desperate measure to get the necessary signatures, Jefferson suggested moving Congress to his house. Finally the delegate and enough of the others arrived to sign the document, but because of further mishaps, the treaty wasn't officially ratified until May, missing the March deadline. Thankfully, Benjamin Franklin was able to get assurances from the British that they would still recognize the treaty despite the lapsed date.

The Congress, which represented the government of the United States in its entirety, soon gained a reputation as a do-nothing assemblage. "Our body was little numerous, but very contentious," Jefferson noted, but he asked how it could be any other way if Congress was doomed to be composed of lawyers "whose trade it is to question everything, yield nothing and talk by the hour."

As a nation, the United States of America continued to flail to keep above water. Conflicts soon arose between the states. Some attempted to tax goods moving across borders; when New York refused to sign a free-trade agreement with Connecticut, Connecticut decided in 1787 not to import any goods from New York for one year. The result was a quasi embargo between the neighboring states.

Many states began issuing their own paper money, which the public had less faith in than the hard metal coins from other nations. There was also the problem that the differing state currencies meant that citizens had to calculate exchange rates every time they crossed state lines. So many types of currency came into circulation that merchants' newspaper advertisements frequently mentioned that they accepted half-joes, crowns, coppers, pistareens, doubloons, guineas, pistoles, or moidores—that is, any coin from Britain, France, or Spain. Not surprisingly, counterfeiting became common.

This slapdash arrangement, combined with conflicts among the states, quickly undermined confidence in the government, and there was the very real possibility that it would implode.

All this mayhem didn't exactly cause the country's economy to chug along as well as it could, and a financial depression began causing many farmers to lose their land to creditors. In Massachusetts, where the rebellion against King George had begun, another conflict was brewing. In the uprising, which became known as Shay's Rebellion, a group of about a thousand armed farmers began marching on Boston hoping to overthrow the government there, but they were beaten back by a mercenary army assembled by Boston merchants.

The Articles also brought financial distress to the federal government. Because they forbade Congress from imposing taxes (the funding of the federal government depended on payments by the states, payments that weren't always quick in coming), the government was nearly bankrupt. It was not even able to pay the interest on the national debt. Amendments to the Articles, introduced in 1781 and again in 1786, would have allowed the federal government to impose taxes, but both times the measures were stopped by the rule that required the unanimous consent of the states to modify the Articles.

When the federal government lacked the money to pay the army, real problems began. Some of the military officers began talking openly of mutiny, but when Congress offered half pay, the men decided that that was the best deal they could get, and they dropped their demands—except for a few men, who were so angry that they began planning a full coup d'état, hoping to install a military government in place of the Congress. It is the only known case of a planned military coup in United States history. A shaken George Washington, who was appalled that his former officers were about to throw "themselves into a gulph of Civil horror," began to worry that the government was about to fail.

"What astonishing changes a few years are capable of producing," Washington said. "From the high ground we stood upon from the plain path which invited our footsteps, to be so fallen! so lost! it is really mortifying."

George Washington was often a pessimist, but in this case his sense of dread seemed justified. Dr. Benjamin Rush of Philadelphia said that the American citizenry was about to "degenerate into savages capable of devouring each other like beasts of prey."

After the threat of a military coup was over, Washington became concerned that western settlers would form an allegiance with some other country—"the touch of a feather would almost incline them away," he said. Spain was certainly interested in adding land to its Florida or Louisiana territories.

The Spanish governor in New Orleans paid bribes to Daniel Boone, who began trying to drum up pro-Spanish feelings in Kentucky; in Tennessee, officials named a territory after a Spanish official, hoping to win favor with the Spanish government.

Meanwhile, in the Midwest, the former hero of the Revolutionary War, George Rogers Clark, had to give up a campaign to defend western settlers against Indian attacks because he ran out of supplies. An angry Clark said, "No property or person is safe under a government so weak as that of the United States."

Finally there was enough concern among the leaders at both the state and national levels that in 1787, the Confederation Congress authorized a convention to revise the Articles. It was a difficult—and possibly illegal—assignment. Only nine of thirteen states authorized the convention, and by the laws of the Articles of Confederation themselves, any changes made to the Articles had to have the approval of all of the states. Several of the states had sent delegates with marching orders to do whatever necessary to fix the situation, even if it meant a new constitution. Twelve of the thirteen states sent representatives (only Rhode Island declined to participate), and through the summer of 1787 the delegates, after much arduous political debate, designed the first democracy. The new Constitution provided for executive, legislative, and judicial branches of government, and it was submitted to the states for approval on September 17, 1787.

By spring of 1788 it looked as if the new Constitution was going to be voted down. An organized group called antifederalists were opposed to the Constitution because it said nothing about individual rights. The group included such well-known figures as Patrick Henry; Richard Henry Lee; John

Hancock; Sam Adams; two future vice presidents, Elbridge Gerry and George Clinton; and future president James Monroe. They were able to block ratification in several states, but when the Bill of Rights was introduced, their opposition to the Constitution melted away.

The U.S. Constitution was finally approved by the last state, Rhode Island, on May 29, 1788, and the next fall voters went to the polls to vote for the first time for a U.S. president. The next year, on April 30, 1789, George Washington took office as the first president of the United States, and the Confederation of the United States was finally, completely at an end.

WHAT IF?
The Articles of Confederation had remained the basis of American government?

With the primacy of states' rights in the confederacy of the United States, the states began focusing more and more on their own needs. Over time, this self-interest exacerbated sectional crises, causing states to make and break alliances as issues dictated. The hodgepodge nation struggled along until 1851, when a large army invaded the young nation from the north, and the states found they had to fight for their survival.

The most crippling of problems arose almost at once, when three states—Kentucky, Maryland, and Missouri— decided that they would not join in the battle and declared themselves neutral. This put the country at an immediate disadvantage. Not only were the men of these states not available for service, but perhaps more important, early in the fighting the industrial facilities of these states weren't available for the effort. In the end, decisions by these three states to remain neutral didn't keep them out of the conflict, because almost immediately armies were marching across their countryside.

Because in the Confederation each state was autonomous, there was no national effort to raise troops—each state was solely responsible for persuading men to fight—and the politicians from each state were allowed to determine where the men would be sent to fight and who the commanders would be. This was a confusing situation even when it worked as well as it could, which wasn't often. Also, because there was no truly national war effort, raising a navy was difficult. Within weeks after the war began, the enemy was able to establish a hostile blockade offshore. On the battlefield the situation was even more difficult. It was not unheard of for men from one state to refuse to fight under the command of an officer from another state, and when troops were needed to replenish the ranks, senior officers often had to negotiate with local politicians to get the men that they needed.

The situation worsened when the governor of Alabama decided the state needed its own army—one separate from the federal army—to defend its shores in case it was attacked along its very short coast. Quickly the other states decided that they needed their own armies, too, and it was common for each state to hold twenty- to forty thousand men out of the army for its own militia. As one governor became increasingly paranoid about the prospect of a coastal invasion, he insisted that even the regular army troops from his state be recalled from the front lines and returned to defend against the phantom invaders. Another governor said that he had allowed the soldiers to leave his state only while a national emergency existed, but that because there was no emergency (despite the fact that the country was at war), the soldiers should be released to come home to tend to their crops. Still another governor withdrew his troops from the federal army in the middle of a battle.

Quickly state militias became the most popular form of military service. Not only did the men get to remain close to home, but discipline in most of the state militias was very lax. Men were allowed to come and go as they pleased, and one observer said that they were "little better than an armed mob." So many men were given appointments as militia officers—a comparatively easier job than a regular army soldier—that one general cracked that a typical militia unit was composed of "three field officers, four staff officers, ten captains, thirty lieutenants, and one private with misery in his bowels."

Each state was also responsible for equipping its own soldiers. After one state purchased a large shipment of arms, which included a number of new brass cannons, it refused to let the shiny new weapons leave the state. Many states refused to send arms with their troops who were marching off to the battlefield, insisting that the weapons were needed at home in case the enemy invaded the coast. Some states refused to send any sort of arms out of the state at all; one governor went so far as to issue an order that any officer who allowed his men to carry arms out of the state would be imprisoned.

The result was that although there were men volunteering for the war effort, they had to be turned away because there were not enough weapons to go around. Eventually the army was able to assemble a force of 250,000 men by the end of the first year of the conflict, but the secretary of war reported to the president that an additional 350,000 men could have been sent to the battlefield if the states had released the weapons that they held in their warehouses. One general complained that the stubborn governor of Texas had done more to disrupt the war effort by "his factious opposition to the laws of Congress on such grounds as the 'dignity of state' . . . and the extreme state-rights' construction of the Laws of Congress" than had the enemy.

Besides weapons and ammunition, each state was also responsible for equipping its soldiers with soft goods such as uniforms, tents, and blankets. This resulted in the many states competing against each other in the marketplace, driving up the prices paid for war materiel. Alabama continued its rebellious stance by charging the government $250,000 for shoes, taking a $100,000 profit for the state coffers. North Carolina refused to contribute more soft goods to the effort than were needed to equip its own soldiers. As a result, men from other states were dying from exposure to the harsh winter conditions while new uniforms and blankets sat unused in a warehouse in the state. Worse, when foreign governments saw that the states could not even trust each other to purchase uniforms, they refused to extend credit to the states to purchase goods overseas.

Surprisingly enough, the war against the invaders achieved early success. The rank-and-file soldiers were patriotic men determined to save their homeland. But the command was rotten at the top: As the war bogged down, some of the top leaders, the nation's elite, began withdrawing from the effort. Not long after, there were voices in Congress that implied mutiny on the battlefield and secession in the government. The attacks were personal as well, and soon the president was a virtual prisoner in the nation's capital, unable to travel for fear that his authority would be overthrown and unwilling to entertain socially because he was so despised that any social function might itself dissolve into violence.

In hindsight, the result of this selfish behavior of the states and their insistence on their individual sovereignty doomed the Confederation. After four years of fighting the nation was forced to surrender.

An astute reader of American history will recognize this alternate history, because this is a true story. Although set twenty years earlier, all of the above events did occur, in the nation of the Confederate States of America during the Civil War. The Confederacy's slavish adherence to states' rights doomed their government to failure. If the states had supplied all of the men and materiel in their control, the Confederate army would have had six hundred thousand in the field in 1861, and might have had a chance to seize a victory. Instead, fewer than half that many soldiers went into battle because of the nation's self-imposed limitations.

If the early leaders of the first government of the United States had similarly insisted on preserving the Articles of Confederation, the United States might have been as ill-equipped to meet a crisis as the Confederate States of America was eighty years later.

Chapter 5

The Sword over the President

❧

The Twelfth Amendment to the Constitution ensured that the vice president would run on the same ticket with the president for the executive offices. What if the original procedure had been left in place?

UNDER the original plan spelled out in the Constitution, no one campaigned for the vice presidency. There was an election for the presidency, and the person who came in second received the vice presidency as a consolation prize. It seemed like a good idea at the time.

This plan guaranteed that someone reasonably competent would assume the presidency if tragedy struck, and the newly elevated president would have something of a mandate when he assumed the office, because a large portion of the citizenry would have voted for him in the previous election.

What the Constitution committee chose to ignore was the influence of factions, or political parties, and how they might play into the two-tier executive branch of government. President George Washington and Vice President John Adams had both been Federalists, but when Thomas Jefferson became Adams's vice president following the election of 1796, there was a problem.

Jefferson belonged to a party that called itself the Democratic-Republicans, the precursor to today's Democratic Party. With the Federalist Adams in the presidency and Democratic-Republican Jefferson in the vice-presidency, plenty of material existed for a powdered-wig situation comedy. First Jefferson refused to attend cabinet meetings, which led to the unfortunate precedent for our nation that veeps weren't often welcome to attend the meetings in subsequent administrations. Then Jefferson refused President Adams's request to travel to France to help settle a spat that threatened to develop into a full-blown war, even though Jefferson had been a popular ambassador to that country. Jefferson further opposed his president by helping to fund an anti-Adams newspaper, which was illegal because the era's sedition laws in effect made complaining about the government a federal crime. After the newspaper editorialized that the coming election would bring a choice between "Adams, war and beggary, and Jefferson, peace and competency," Jefferson wrote the editor to cheer that "such papers cannot fail to produce the best efforts." Vice President Jefferson was not going to win any teamwork awards.

In the election of 1800, for the only time in United States history, the incumbent vice president defeated the incumbent president in the race for the nation's top office, when Jefferson completed his assault on Adams by beating him at the polls. But Jefferson couldn't laugh about how he had undermined the Adams administration, because he quickly found out that his vice president was going to be considerably more trouble than even he had been.

The people who worried about such things in 1800 thought that in the upcoming election they had all of the problems created by Jefferson worked out. Because the vice president would be the person who received the second highest number of votes in the electoral college, and not in the general election, Jefferson's Democratic-Republicans decided that they would have some presidential electors vote for Jefferson and some presidential electors vote for their pick for vice president, Aaron Burr. This way they would have men from the same party in the top two offices, and they could avoid the type of harmful undermining of support that an opposition vice president could orchestrate.

Somehow, either by poor planning or by the machinations of Burr, both Jefferson and Burr received 73 electoral college votes for president, and the outcome of the election had to be decided by the House of Representatives. Burr could have simply said that he was interested only in the vice presidency, which would have allowed Jefferson to step easily into the presidency, but he didn't do that. Whether Burr had planned the whole mess beforehand, or whether he was trying to take advantage of fortuitous circumstances, isn't clear. But Burr knew an opportunity when he saw one, and once the election was thrown to the House of Representatives, he began lobbying for the congressmen to pick him, and not Jefferson, as president.

Burr's plan almost worked. When the House voted in February, the winner needed only to carry a majority of the sixteen states to win the presidency. On the first ballot Jefferson received eight votes, Burr six, and two states were so undecided that they abstained. The ordeal went on for thirty-six ballots over a span of six days, until finally some of Burr's supporters gave up and Jefferson won the office. Not surprisingly, President Jefferson didn't consult Vice President Burr on many decisions.

After the disastrous and potentially dangerous vice presidencies of Jefferson and Burr (besides the electoral shenanigans, Burr had killed Alexander Hamilton in a duel while vice president), it became obvious that a change had to be made. Bills for proposed constitutional amendments were introduced into both the House and the Senate that would have simply done away with the vice presidency. When neither of these bills could gain the necessary support, another bill was introduced that directed the electoral college voters to specifically cast their votes for one candidate for president and one candidate for vice president. If either candidate failed to get the necessary electoral college votes, the House of Representatives would decide who was to be president, and the Senate would decide who was vice president. (It might seem that the House should vote on the vice president and the more deliberative Senate pick the president, but that's not the way they figured it out.) This amendment sailed through Congress and in 1804 was quickly adopted by the states as the Twelfth Amendment to the Constitution—only the second change to the Constitution after the Bill of Rights, and the only real structural change made to our federal government's organization since the Constitution.

WHAT IF?

But what would the nation's executive offices have looked like had the original constitutional plan for the vice presidency continued?

After the disaster of Aaron Burr's vice presidency, the nation was eager to have a less interesting vice president. Finding someone less colorful than the murderous, treasonous, adulterous Burr was easy. But De Witt Clinton, Thomas Jefferson's second vice president, epitomized the opposite extreme. Not an allegation or even a rumor was spoken about him, which was just fine with most people. For the next thirty-five years the vice presidency was barely visible.

More problems in the unfortunate history of the vice presidency came soon enough. In the election of 1840, Martin Van Buren's own vice president, William Henry Harrison, defeated him—and then Harrison died just one month after his inauguration, giving the presidency back to Van Buren. Van Buren had been nicknamed the Little Magician for his ability to make political opponents disappear during the Jackson administration. (He had convinced Jackson's entire cabinet to offer their resignations, which elevated Van Buren's power in that administration.) With Van Buren's reputation for political trickery, some less reasonable members of society publicly accused Van Buren of murdering Harrison. Most others weren't convinced—for one thing, if the accusations were true, it would have been the first time that the murder weapon was pneumonia. But the suspicions dogged Van Buren for the remainder of his term.

The history of the vice presidency took a more serious turn twenty years later in the election of 1860. After a difficult and contentious election that saw each major party split into Northern and Southern wings, Republican Abraham Lincoln won eighteen states. Except for the western states of California and Oregon, all of the states that went for Lincoln were north of the Mason-Dixon line. Finishing second in the four-way presidential race was John Breckinridge, a senator from Kentucky, who had been able to win most of the slave states. The nation had split into two blocks of voters for the second election in a row, and it was obvious to many Southerners that from now on they could aspire to place a candidate no higher than vice president, while the North would always vote for an antislavery candidate. Having no voice in national affairs, the Southerners realized, was as bad as not being in the nation at all.

Within days after Lincoln took office, a cascade of Southern states announced that they were leaving the Union. Washington, D.C., was bound by the Confederate state of Virginia on one side and by Maryland on the other. Maryland had not joined the Confederacy, but the newly inaugurated president wasn't beloved in that state either. Of the ninety thousand Maryland voters who went to the polls in 1860, only a little more than two thousand had voted for Lincoln and his antislavery views. Washington, D.C., was an island in a hostile region for the abolitionist president, and although it shocked the nation at the time, through the lens of history the result of this isolation is no surprise.

Just three weeks into his term, Lincoln stood on the front lawn of the White House on an unusually pleasant March afternoon greeting well-wishers. One of the well-wishers approached the president with a coat draped over his arm, and when Lincoln held out his hand, the man fired the pistol that was hidden under the cloth.

Lincoln immediately lost consciousness and lay in a coma in the White House as aides and the nation awaited his death. The gunman had been able to dart out of the crowd in the confusion, and as the days passed with Lincoln lingering near death, no one was arrested for the attempted murder. The citizens were outraged. There was a strong need to blame someone for the attack on the president, but no suspect could be found. Obviously, public opinion said, the gunman had been a

Southern sympathizer who had committed the act hoping to make the pro-slavery Breckinridge president. But could Breckinridge himself have been involved?

Breckinridge, charismatic and good-looking, was only forty years old when he claimed the vice presidency. He had risen in politics because he had been coy about where he stood on the slavery issue, but Stephen Douglas, another candidate in the crowded field in 1860, noted, "Breckinridge may not be for disunion, but all of the disunionists are for Breckinridge." How much the disunionists were for Breckinridge soon became a crucial national question.

Accusations were soon directed at the vice president, and the suspicions and the overwhelming opposition to his policies in the North prevented Breckinridge from assuming the presidency from the completely disabled Lincoln. "Mr. Breckinridge should not try to usurp the now-vacant office of the Presidency until Providence decides our fate," declared William Seward, the secretary of state, summing up the opinions of the others in the Lincoln cabinet.

A week after the attempted assasination of President Lincoln, Breckinridge committed what was either a desperate or a foolish act. A drunken mob of a dozen men had made its way to the door of Breckinridge's hotel room late at night, and Breckinridge was only able to keep the men from busting the lock and breaking down the door to his room by shoving a chest of drawers and other furniture against the door. Washington police soon rousted the men out of the hotel, but the vice president realized that Washington was becoming even more dangerous for him than it had been for Lincoln.

The next night, Breckinridge was stopped by a guard

as he tried to flee across the Potomac River into Confederate Virginia. When the vice president gave the young corporal guarding the crossing a fake name and tried to keep his hat pulled down over his eyes, the corporal recognized him and placed him under arrest. Whether Breckinridge was simply trying to escape from a hostile Washington or whether he was planning to take his place as one of the leaders of the Confederacy will never be known. The vice president was placed in a Washington jail, while Attorney General Edward Bates said that the charges against Breckinridge were "substantial, but secret." The flimsy evidence against Breckinridge didn't stop most people in the North from assuming his guilt. Breckinridge's attempt to disguise himself and escape was evidence enough of his participation in a conspiracy to kill the president, most people thought.

A month after being arrested, Breckinridge was brought to trial in a military court, charged as an accomplice to the attempted murder of Lincoln, despite the fact that the assassin still had not been found.

The trial of Vice President Breckinridge was a bleak moment in American justice. There was little evidence available in the crime. Lincoln still lay in a coma in the White House, the shooter had not been found (already in Havana, was the conventional wisdom), and no one had stepped forward with even hearsay evidence of Breckinridge's desire to see Lincoln killed. However, the vice president had tried to pass through military lines to Virginia, obviously hoping to join the Confederacy. And that was enough evidence of treason for any Union jury. "Breckinridge should die the most disgraceful death known to our civilization—death on the Gallows," opined the *New York Times*, and in May 1861, Vice President Breckinridge was sentenced to death. After several hurried and unwelcome appeals, in September 1861, while President Lincoln still lay in a coma, Breckinridge was hanged for his alleged role in the shooting of the president.

Although the events surrounding the vice presidency were never again as violent as they were during the days of the Civil War, the vice presidency was a constant threat to any president's ability to govern. When President Woodrow Wilson suffered a debilitating stroke in 1919, the Republicans worked vigorously to install their man, Vice President Charles Evans Hughes, in the White House, going so far as to challenge the president's authority before the Supreme Court. In 1946, after Thomas Dewey had assumed the presidency following the death of Franklin Roosevelt, Democrats in Congress successfully blocked every legislative effort Dewey put forward. As a result of the public outrage at these tactics, the Democrats lost control of Congress in 1946 and weren't able to regain a majority in Congress for sixteen years.

From the Dewey presidency on, each president who had to deal with the opposing party in control of Congress faced impeachment charges, and several presidents were forced to withstand actual impeachment proceedings. None of the impeachment attempts succeeded in removing the president and installing the party's own vice president, but that may not have always been the point. Embattled presidents were nevertheless forced to spend months or even years fighting off challenges to their office.

The closest a president came to being removed was in 1973, following the reelection of Richard Nixon. Impeachment proceedings had begun over the secret bombing of Cambodia and Laos during the Vietnam War, but they picked up speed when news of the scandals known collectively as Watergate began making headlines. For once it appeared that the opposing party had sufficient evidence to remove the president and install its own man as the nation's chief executive officer. Vice President George McGovern was so confident of his upcoming promotion, in fact, that he began leaking to the press the names of those whom he would appoint to his cabinet.

But Nixon was a skillful politician. Because his impeachment and removal would cause such a dramatic shift in the nation's leadership—country club conservatives and ex-military officers would be out, while liberal college professors and scruffy war protestors would be in—Nixon was able to convince many in the country's heartland that the impeachment hearings weren't about his misdeeds at all. Instead, the debate was an attempt by the Democratic Party (or the "Commicrat Party of Betrayal," as one Nixon White House aide phrased it) to elevate McGovern to the presidency. It was nothing less than a left-wing political coup d'état, many in the nation believed, and the effort to remove the president had little public support. After a dramatic Senate showdown, Nixon was able to serve out his term until 1976.

The consolation-prize vice presidency ensured that a capable person could assume the presidency if needed, a person whom a large portion of the electorate had actually preferred in that position, giving the new president an immediate mandate. And there was little doubt that the nation benefited from the overall confidence in its leaders. But the office had few enthusiastic fans outside of political cartoonists and stand-up comics. (What better material could there be than the bitter Jimmy Carter sulking around the optimistic Reagan White House or the acerbic Bob Dole offering commentary about the scandal-scarred

Clinton administration?) Unfortunately, the day-to-day reality of the arrangement meant that every president would include in his administration one high-ranking critic from the opposite party, a measure that literally threatened some presidential administrations and damaged the credibility of even the strongest administrations. The American vice presidency was truly the world's most peculiar high office.

Of course, our history didn't turn out that way. The Twelfth Amendment to the Constitution allowed presidential candidates to bring along vice presidents from their own party, eliminating the possibility of political tricks or outright violence that may have become associated with the nation's second office. Plenty of people thought that the amendment was a bad idea. One senator said that the office would attract only men "of moderate talents, whose ambition is bounded by that office." In other words, the vice presidency would become a high political office for people who wanted to *be* something important, not for people who wanted to *do* something important, an overweening characteristic that could be described as "bland ambition." Another senator described the office as being as useful to government as "a fifth wheel on a coach."

Because of the vice presidency, our nation has been one fragile human life away from such improbable presidents as Richard Mentor Johnson, Hannibal Hamlin, Spiro Agnew, Walter Mondale, and Dan Quayle—men little known or of questionable capabilities. Thanks to the vice presidency, Chester Arthur, a man Woodrow Wilson accurately described as "a nonentity with sidewhiskers," was able to become our nation's most unlikely president after the death of James Garfield.

Chapter 6

The Prophet's Victory

Tecumseh planned to create a new nation of Native Americans, but a decision by his brother, the Prophet, against Tecumseh's wishes brought an end to this vision. What if the Prophet had waited for Tecumseh?

THE Shawnee chief Tecumseh had a grand plan. For 250 years, since the Europeans had first arrived in America, they had fought and stolen from the Native Americans. But Tecumseh knew that he could stop the taking of Indian lands if the many Native American tribes would all band together as one.

Tecumseh planned to create a political system similar to that of the Europeans, a great confederation called the Union of the Fires. This nation of Native Americans would be recognized and respected by the white people and backed with the fighting forces of all the Eastern forest tribes. The fighting force would be a necessary part of this plan, because Tecumseh's vision had a darker side, too: He wanted to kill every American west of the Appalachian Mountains, every man, woman, and child, and force the Americans to abandon the Mississippi Valley.

The Native Americans who had inhabited the forests of North America had been pushed west to the edge of the prairie, and their cultures were in disarray. The white settlers tried to get the Indians to take up farming, but the Native American males thought that farming was too feminine an activity for men. Instead many of them spent their time drinking and fighting. William Henry Harrison, the governor of the Indiana Territory, described the situation: "Killing each other has become so customary amongst them that it is no longer a crime to murder those whom they have been most accustomed to esteem and regard. Their chiefs and their nearest relations fall under the strokes of their tomahawks and knives."

White settlers were pouring into the Indiana Territory from Kentucky and Tennessee in the south and from Pennsylvania in the east, and as quickly as he could, Harrison negotiated treaties with the Indian chiefs to purchase portions of their land. After meeting with Harrison, the Delaware, Shawnee, Potawatomi, Miami, Wea, Kickapoo, Piankashaw, and Kaskaskia had agreed to give up most of what is now Indiana and Illinois. Harrison recognized that many of the Indians didn't understand the treaties: "Now if a poor Indian attempts to take a little bark from a tree, to cover him from the rain," he observed, "up comes a white man and threatens to shoot him, claiming the tree as his own." Rather than being sympathetic to this interpretation of property rights, though, Harrison found it amusingly naïve and continued to press for treaties that would move the Indians off their homelands.

Although many Native Americans didn't understand that the treaties required them to leave the land, Tecumseh did, and so he bitterly opposed the treaties. Most Native Americans who had lived for centuries in the forested Eastern portion of the United States resided in tribes of fifty to one hundred people, called fires by the Native Americans. Other chiefs, including the great chief Pontiac, had envisioned pulling together to oppose the advancing white Americans. But Tecumseh added an important ingredient to the vision. Tecumseh made a new religion a part of his plan, and what could have been just a political idea was transformed into a spiritual quest. To make his vision a reality, Tecumseh relied on the talents of his brother, who would bring the idea both initial success and eventual failure.

Tecumseh's brother Laulewasika, or "the Noisemaker," was much younger than the famous warrior. Their father had died while Laulewasika was a small child, and unlike Tecumseh, Laulewasika had no one to teach him how to hunt and fight. Laulewasika was so incompetent with weapons, in fact, that he permanently damaged his right eye trying to shoot an arrow. When he was a young man he tried to make himself into a medicine man, but he was a failure at this and became known only as a drunk, bragging nuisance.

That changed after a bizarre episode. One night in 1805, when he was about thirty years old, Laulewasika passed out in what seemed to be an alcoholic stupor. It soon appeared, however, that he had stopped breathing, and as his wife grieved, preparations were made for his funeral. Before the burial service began, however, Laulewasika suddenly awoke.

He told the startled Shawnee that, while he was unconscious, he had seen a vision from a supreme being whom he called the Master of Life. In his vision, he had seen the afterlife divided into two worlds. One was a paradise where Indians were allowed to hunt and fish in an abundant land, where there was plenty of game and where the cornfields were fertile and productive. But Indians who did not live good lives were sent into a type of purgatory, a large hut where fires burned all of the time. The worst of the Indians were burned immediately, and others were subjected to fiery torture until they had paid for their sins. Only then were they allowed to enter the paradise, but even so, they could not completely enjoy all of its pleasures.

After his vision, Laulewasika took the name Tenskwatawa, the Prophet, and told the Shawnee that they must give up their desire to accumulate wealth and strive to live in accordance with the desires of the Master of Life. He instructed them to reject many of the trappings of the Europeans, specifically enjoining them not to eat European food, such as pork and bread, and not to use guns or wear European clothes. He also told them to treat each other with affection and to end the violence among themselves; they should live in monogamous relationships and avoid promiscuity, and pray to the Master of Life twice a day. Further, he commanded his followers to give up alcohol—if they imbibed, the tribal punishment would be torture.

Finally, he instructed his followers to avoid the Americans, who were created differently from the French, British, and Spaniards. The Americans, he said, were created not by the Master of Life but by the Evil Spirit, and had sprung up "from the scum of the great Water when it was troubled by the Evil Spirit. And the froth was driven into the Woods by a strong east wind . . . They have taken away your lands, which were not made for them."

The vision of the Prophet became well known among the neighboring Shawnee and Delaware tribes, and he attracted many followers. Although he encouraged a spiritual life, he strongly relied on violence to enforce his teachings. Members of his tribe who continued to associate with Americans were executed for their sins. The Prophet and Tecumseh also increased their violence against Americans. After a group of Tecumseh's warriors acting on behalf of the Prophet brutally killed a missionary, the other missionaries in the area reported back to their superiors that they were "ready and willing to be used in further service of the dear Savior, *only not here.*"

The Prophet had his doubters, and among them was Harrison. Alarmed at the increasing violence against both white settlers and amongst the Indians themselves, Harrison sent a warning to the Shawnee and Delaware tribes: "The dark, crooked, and thorny [road] you are now pursuing will lead to woe and misery. But who is this pretended prophet who dares to speak in the name of the great Creator?" Harrison told them: "Examine him. If he is really a prophet, ask of him to cause the sun to stand still; the moon to alter its course; the rivers to cease to flow; or the dead to rise from their graves. If he does these things, you may then believe that he has been sent from God."

With his challenge, Harrison unwittingly added to the Prophet's power. In June the Prophet told his followers that he would cause a Black Sun, which the Shawnee called Muku-taaweethee Keesohtoa, a harbinger of war. As Indians from nearby tribes assembled in Greenville, Ohio, to see if the Prophet could actually make the sun go black, the Prophet disappeared into his lodge and did not come out for several days. Then, on June 16, a full eclipse occurred, and the Prophet had his miracle.

Some American settlers noted that a team of scientists had traveled through the Ohio Valley that summer to study the eclipse that they knew was imminent, and it appears likely that the Prophet knew of these men and their purpose for coming to the frontier. Regardless of how the miracle was performed, it had the desired effect, and the Prophet's reputation spread throughout the Old Northwest.

Near where the Tippecanoe River flows into the Wabash River in present-day Indiana, Tecumseh and the Prophet built a new, large village that attracted people from all tribes. Like

Tecumseh's political structure and the Prophet's religion, the village, Prophetstown, incorporated a mix of Native American and European cultures. Unlike other Indian villages, Prophetstown was laid out along a main street, and it included several log cabins, which had not previously been used by Native Americans, but had been adopted because of their greater strength and warmth. The village also included bark wigwams, as well as tepees built by the Winnebagos and Kickapoo tribes. Besides these dwellings, the village also boasted a granary that was raised on stilts; a House of Strangers, a type of hotel for incoming families; a blacksmith shop; a council house; and the Prophet's octagonal medicine lodge.

Soon the chiefs and warriors of the Kickapoo, Ottawa, Potawatomie, Winnebago, and Wyandot tribes all proclaimed Tecumseh as their chief. In the summer of 1811, Harrison decided that he needed to press Tecumseh for a peace treaty, but the meeting went badly, with Tecumseh demanding that the white settlers leave the area. After the meeting Harrison was convinced that only a military attack would settle the dispute with Tecumseh's forces, and he requested and received an additional regiment of soldiers from Pennsylvania.

Tecumseh wanted to expand his Union of the Fires beyond the Indiana Territory, and in late summer he traveled south to ask other tribes to join his confederacy. Telling the southern Creeks that he had traveled "more than a thousand miles, from the borders of Canada," to deliver his message, he persuaded many of the Creeks, Muskogees, and, in Florida, the Seminoles to join with him.

In late September, Harrison decided that he had to press Native Americans at Prophetstown while Tecumseh was away on his journey and while the weather still allowed his army to move. He told the townspeople of Vincennes, the capital of the Indiana Territory, that he and his army might be killed and, if so, the people should fortify the courthouse, assemble there, and send for help from the governor of Kentucky. Harrison then mustered his 210 soldiers, plus another 600 volunteers from the region, and began the 150-mile march to Prophetstown.

Harrison hoped that the sight of the large force of men would convince the Native Americans to sign a peace treaty. When he arrived at Prophetstown, he sent a patrol of men under a flag of truce into the village, but the Prophet refused to meet with them. However, that night three warriors from Prophetstown approached Harrison's camp and said that the Prophet would meet with Harrison in person the next day to work out a peace treaty.

Tecumseh had warned the Prophet not to engage the Americans in a battle until he returned, and the British, who were supplying Tecumseh's warriors with weapons, had told the Indians not to fight until the British were ready to support them. But the Prophet decided that this was the time for war, and that he would be the one to lead the tribes into battle. Instead of meeting with Harrison to discuss peace, as he had said, he ordered his warriors to attack.

That night, Harrison warned his men to sleep with loaded weapons, and his instincts were correct. Just after four o'clock in the morning, the guard came running into the camp followed by the howling warriors. Harrison's troops quickly formed ranks and began firing back. The Native Americans didn't fire from behind trees and rocks, which was their normal way, but instead rushed right up to Harrison's lines and fired. The Prophet, who spent the battle sitting on a rock chanting war songs, had told his warriors that he had given them special protection that would cause the white man's bullets to bounce off them. The results were predictably unfortunate. One Winnebago chief needed to repair the flint on his rifle, so he walked up to an American campfire where the light was better. Several American soldiers fired at him, and he fell face first into the fire. When the battle was over, sixty-two American soldiers had been killed, and the bodies of thirty-eight Native Americans were found on the battlefield. However, because the Native Americans always removed the bodies of fallen warriors from the field of battle, Harrison's troops guessed that anywhere from 350 to as many as 1,000 of the Prophet's warriors had lost their lives. After this defeat, the other Indians could plainly see that the Prophet had no special powers, and the followers left to return to their tribes.

WHAT IF?

What if the Prophet had waited for his brother Tecumseh to return before attacking the Americans?

Following his tendentious meeting with Harrison, in which Tecumseh insisted that all white settlers leave the territory, Tecumseh warned his brother that Harrison might launch an attack and emphasized that they must preserve the village. Tecumseh left to meet with the fires to the south, and soon scouts from the village told the Prophet that a large force of armed Americans was traveling by land along the Wabash River. The time to deal with the Americans would arrive soon. The Prophet sent a message to Harrison: "The Great Spirit has told us to build a new home here on the Wabash. I welcome you to join us in our home to see that we only want peace."

The Indians who delivered the message promised that the army would not be attacked along the way, and offered to show Harrison the best route and the best places for the Americans to build their camps. As the army set up camp outside Prophetstown, Harrison sent a patrol of men to the Prophet to discuss peace. The Prophet told the men that although the army was not welcome in the village, Harrison and his officers could meet with him in the council house to discuss peace. Harrison arrived the next day and told the Prophet, "Our blue coats are more numerous than you can count, and our hunting shirts are like the leaves of the forest or the grains of the sands on the Wabash." The Prophet understood too well how numerous the Americans were, but he was not yet ready to provoke a battle.

He responded, "If you cross the boundary of your present settlement it will be very hard and produce great troubles among us. But if you allow us to live here in peace we will not cross into your lands," referring to the land east of the Appalachian Mountains.

The Prophet continued. "Our chief, Tecumseh, has traveled great distances to invite other chiefs to join us. If you attack us now and force us from our home, the other chiefs will rise up against you."

Harrison knew that with the American army in the northern part of the territory, the fort at Vincennes was poorly defended, and he had no doubt that what the Prophet was telling him was true. Tecumseh and his brother had shown great skill in getting the other chiefs to join their Union of the Fires, and as long as the great fighter Tecumseh was still away, there would be less chance that Harrison could establish a lasting peace. Attacking the village now would be the same as Tecumseh attacking Vincennes while he was away, and it would be sure to inflame widespread Indian attacks.

Harrison told the Prophet that as long as there were

no attacks on the Americans, he would allow the tribes to remain in their village, and Harrison and his army began to march back to Vincennes.

As soon as Tecumseh returned to the Indiana Territory, however, the attacks began again. Harrison didn't know whether the Prophet had lied to him or whether Tecumseh, who wanted to kill all Americans, was now leading the renewed attacks. It didn't matter. Harrison knew that the time had come for a battle.

As soon as warmer weather arrived the next March, Harrison assembled the army again and set off for Prophetstown. Remembering the Prophet's warning of an attack on Vincennes, though, this time Harrison left fifty regular soldiers and two hundred volunteers to protect the families there.

The march back to Prophetstown was difficult. The trails were muddy, the fields were flooded, and many of the creeks and rivers were swollen and moving swiftly. As the army approached to within twenty miles of Prophetstown, they came under sporadic attack. Indian warriors, shooting from behind trees, surprised the army as they slogged through the wet trail. By the time Harrison's army had formed ranks and put powder in their guns, the Indians were gone. After a day or two there would be another attack, each time with Tecumseh's warriors retreating before the Americans could launch a full engagement.

Harrison's troops were demoralized, and they were worried about their families in Vincennes. They had good reason to be—Tecumseh had organized a war party that had overwhelmed the soldiers there, killed the townspeople, and burned the settlement.

As Harrison and his army approached Prophetstown, Tecumseh's warriors met them. Tecumseh had assembled more than twelve hundred warriors, and the British had supplied them with rifles and other supplies. Harrison's army, which numbered only six hundred exhausted men, was soon pushed back against the flooded Wabash River, which they were unable to cross. All of the Americans, including Harrison, were killed.

After the battle, Tecumseh attacked and destroyed white settlements north of the Ohio River and east of the Wabash, the land he claimed for his new Union of the Fires. Settlers in Kentucky and Tennessee were terrified at the new Indian confederation, and people in Washington were outraged. "Our people will never be able to sleep as long as this terror is allowed to inhabit a place on this earth," cried Congressman Henry Clay. Together with John Calhoun and other war hawks in Congress, they declared war on Britain over the country's support of Tecumseh and other offenses and began raising an army to attack Tecumseh.

In the summer of 1813, just over one year after the massacre of Harrison's army, an American army of more than four thousand men led by Andrew Jackson, the former senator from Tennessee and the general of the Tennessee militia, marched to the Indiana Territory. The territory had been without a U.S. government representative for more than a year, and many people in Washington were concerned that the British were about to claim it for themselves. Jackson was certainly aware of the political significance of the campaign, but for him and his men, the coming battle was a war of revenge against Tecumseh. Jackson and his men moved quickly through the southern part of the Indiana Territory, destroying each Indian village that they encountered. Tecumseh had assembled an army that included seven hundred British soldiers from

Canada, and the two armies met near the settlement of Terre Haute, midway through the Indiana Territory. The Battle of Terre Haute took place on the high plateau that overlooked the Wabash River, and after two days of brutal fighting, Jackson and the Americans were victorious.

The battle was the first major victory in what would become known as the British-Indian War. Jackson and his militia forced the remaining Indians out of the Indiana Territory and across the Mississippi River. The era of the woodland Indians of North America was over.

After the Prophet led his soldiers to defeat, the tribes that had put their faith in him returned to their homes, and the Prophet fled the area, first going to Ohio and later to a settlement in the plains. Tecumseh returned to find the village of Prophetstown burned and his vision of a great Indian confederacy destroyed. Tecumseh was killed a year later fighting for the British in the War of 1812 in the Battle of Thames. The man who claimed to have fired the gun that killed Tecumseh, Richard Mentor Johnson of Kentucky, became so famous that he was elected vice president in 1836. William Henry Harrison, who became known as Ol' Tippecanoe after the famous battle, was elected president in 1840, but he died just one month after taking the oath of office.

The conflicts between the early American settlers and the Native Americans were brutal affairs in which many on each side hoped to accomplish genocide. With the defeat of Tecumseh's federation at Prophetstown, the vision of a strong nation of Native Americans in the eastern United States had failed.

Chapter 7

Losing the Louisiana Territory

Andrew Jackson won the Battle of New Orleans by folding a band of pirates into his army. What if he had decided to face the British alone?

THE Treaty of Ghent, which ended the War of 1812 between the United States and Britain, was signed in Belgium on Christmas Eve, 1814. Two weeks later, the last battle of the war began.

In the War of 1812 Britain hoped to take back the American states as colonies, and the United States hoped to capture Canada for its own. Both nations failed, and the war was fought to a draw, with no clear victor. The final battle of the war, the Battle of New Orleans, is often thought to be a historical curiosity—a superfluous ending to a largely superfluous war. True, the Battle of New Orleans didn't have much impact on the outcome of the War of 1812. But those who dismiss the battle as inconsequential probably haven't taken the time to study it. The Battle of New Orleans may be one of the most successful—and perhaps important—military victories in American history.

❧

President Thomas Jefferson had understood the importance of New Orleans to the fledgling nation, which is why he had risked violating the Constitution and international treaties in order to purchase it from France in 1803. "There is on the globe one single spot, the possessor of which is our natural and habitual enemy," Jefferson said. "It is New Orleans, through which the produce of three-eighths of our territory must pass to market, and from its fertility [the Louisiana Territories] will long yield more than half of our whole produce and contain more than half of our inhabitants." The Louisiana Purchase doubled the size of the United States and allowed settlers to go as far west as the Rocky Mountains.

But in October 1814, American diplomats meeting in Ghent to work out a treaty learned that Britain had sent a naval armada from Ireland in September with orders to capture New Orleans. The British plan was to take the city and then push north to meet up with British soldiers moving south from Canada, and claim again all of the land west of the Appalachian Mountains.

The Duke of Wellington had just defeated Napoleon in the summer of 1814, so Britain was now able to turn its full fury against the United States. The armada that set sail in November comprised fifty ships carrying fourteen thousand troops. The British were so confident of victory that they also carried supplies to colonize the southern coast of North America. In addition to the soldiers, the ships carried the officers' wives and even a printing press and materials for publishing a newspaper once the British had successfully captured the west-ern half of the United States.

The leader of the American forces in the Southeast was rough-hewn Andrew Jackson, who commanded the Tennessee militia. Although the capture of New Orleans was the British goal, Jackson astutely realized that the British weren't going to risk taking their ships up the Mississippi through the Louisiana delta. Instead of heading to New Orleans, Jackson ordered his men to march from Pensacola, Florida, where they had cap-tured a British fort, to the seaport of Mobile, Alabama. Jackson suspected the British hoped to land in the relative safety of Mobile Bay, and he spurred his men to make the four-hundred-mile march in just eleven days. Jackson had guessed right. The British war plan was to invade Mobile and then move overland to New Orleans and Natchez, Mississippi. Arriving in Mobile, Jackson built fortifications overlooking the bay in just two weeks, which forced the British to reconsider their battle plan.

The British realized that Jackson's preparations in Mobile meant that New Orleans was defenseless, and they quickly set sail for the main object itself. When Jackson's capable collec-tion of spies informed him of this, he also hurried his troops to the Crescent City.

The people of New Orleans were ambivalent about a pos-sible British invasion. They were willing to fly the flag of any government, American, French, Spanish, or British, that would allow them to conduct their business in peace. Rather than working to gain the support of the people of New Orleans, Jackson quickly began making himself disliked by the local citizenry. As soon as he arrived in the city, Jackson imposed martial law, turning New Orleans into an armed camp. When the local legislature complained loudly about this, Jackson threatened to blow the legislature to bits. The citizens were even more upset at rumors that Jackson was willing to go so far as to destroy the city to prevent it from falling to the British. And had the British been able to take New Orleans, Jackson would later claim, "I should have retreated to the city, fired it, and fought the enemy amidst the surrounding flames. I would have destroyed New Orleans . . . and in this way compelled them [the British] to depart from the country."

Jackson also angered the locals when he decided to enlist local free black men as volunteers into his army. He did this not because he was an enlightened individual—he accepted the black soldiers with some displeasure, and he was himself a slave owner—but because he realized that he needed all of the help he could get. The regular troops bristled when Jackson offered the black men full pay, rations, and uniforms, but Jackson angrily replied, "Be pleased to keep to yourself your opinions upon the policy of making payments of the troops without inquiring whether the troops are white, black, or tea."

When even this unusual act didn't raise enough men for the defense of the city, Jackson went a step further: He accepted help from outlaw pirate Jean Laffite.

Seventy miles south of New Orleans was a large protected island that sat in Barataria Bay. It was home to Jean Laffite, his brothers Pierre and Dominique, and hundreds of men who skirted the law by smuggling and "privateering," largely at the expense of the Spanish governments of Florida and Mexico.

When they arrived in Louisiana, the British approached Laffite and offered him the rank of captain and thirty thousand dollars if he would fight with them. Considering that Laffite had warehouses filled with more than 1 million dollars' worth of booty, the British offer wasn't much of a temptation. A letter delivered to Laffite the next day offered further incentive: It was the understanding of the British navy, the letter said, that many British seamen had been pressed into service in Laffite's forces, and it might be necessary for the warship *Sophia* to free these men and destroy Laffite's village at Barataria Bay.

Laffite insisted that he needed two weeks to consider the Brits' generous offer, and he then sent a letter to the New Orleans militia:

> *This point of Louisiana, which I occupy, is of great importance in the present crisis. I tender my services to defend it; and the only reward I ask is that a stop be put to the proscription against me and my adherents, by an act of oblivion for all that has been done hitherto. I am the stray sheep, wishing to return to the sheepfold. . . . Should your answer not be favorable to my ardent desires, I declare to you that I will instantly leave the country. . . .*
>
> J'ai l' Honneur d'être
> M. le Gouverneur
> *J. Laffite*

Some in New Orleans thought the letters were a trick to get Jean Laffite's pirates released from jail, and the local militia responded by sending a gunboat to Laffite's island and arresting eighty more of his men. Jackson then issued a statement to the people of New Orleans: "I ask you, Louisianians, can we place any confidence in the honour of men [the British] who have counted an alliance with pirates and robbers?" the letter asked. "Have they not made offers to the pirates of Barataria to join them, and their holy cause? And have they not dared to insult you by calling on you to associate, as brethren with them, and this hellish bandetti?"

Jackson had patched together an army of two thousand men from his Tennessee volunteers, the local militia, and the freed black men, but the British army was calculated to number some twelve thousand. Most of the soldiers from Kentucky had arrived in New Orleans without guns. Jackson was incredulous when he was told: "I don't believe it. I've never seen a Kentuckian without a gun and a pack of cards and a bottle of whiskey in my life." Jackson decided that he should ignore the concerns of the locals—and his own statement to the people of New Orleans—and discuss the matter with Laffite. Although neither man recorded the conversation that day, it is easy to guess Laffite's arguments. The British had offered him a commission, but he had declined their proposition, which proved his loyalty to the United States. He was willing to offer Jackson his men, his cannons, and a large store of ammunition and flints that he had safely stored in the marshes. In exchange for his cooperation, Laffite asked for amnesty for his men. Jackson agreed.

It must have been quite a sight in New Orleans as the jails were opened and the convicts were handed guns. As word spread to Laffite's men who were hiding out on the islands below the city that they were going to join Jackson's army, the pirates began arriving in New Orleans in groups, ready to fight. In the few days that he was in New Orleans, Jackson had built an army of nearly seven thousand men.

The Battle of New Orleans began on the morning of January 8, 1814, while Jackson was eating breakfast with his officers. As more than a hundred shells and cannonballs hit the house where Jackson was staying, he ran out unhurt, and hurried to the defenses to spur on his men. "Don't mind these rockets," he shouted, waving his hat. "They are mere toys to amuse children."

One of Laffite's men, his brother Dominique You, commanded the pirate artillery, and he instructed his men to fill their cannons with chain shot and any other material they could find that would tear the flesh of the British soldiers. The morning of the attack, Jackson rode to the artillery line and found Dominique and his men casually making coffee. Jackson paused long enough to enjoy a cup himself. A short while later, with the battle under way, he rode by Dominique's artillery line and again found the men relaxing. Shouting from his horse, Jackson asked why they hadn't joined in the battle. Dominique explained that the gunpowder he had been supplied was of poor quality. "It is fit only to shoot blackbirds with, not Redcoats," he explained. Jackson, with typical fury, told his officers that if the pirate band wasn't supplied with quality powder, he would have the ordnance officer shot. Soon the pirate cannons were blazing.

As the cannons on both sides continued playing their prelude, the British infantry charge began. Marching to drums, the soldiers approached the silent American lines. When the redcoats were one hundred yards from the American fortifications, a local New Orleans band behind the lines began to play "Yankee Doodle," and the American soldiers opened fire. A Frenchman reporting on the battle later explained, "When these Americans go into battle they forget that they are not hunting deer or shooting turkeys and try never to throw away a shot."

The British had charged the American line with 7,500 soldiers, and within minutes they had suffered more than 2,000 casualties and had nearly 500 more soldiers taken as prisoners. The Americans had lost just thirteen men, and many of these casualties occurred when the men were shot as they walked through the battlefield by British soldiers who did not realize that their side had already surrendered.

WHAT IF?
What if Andrew Jackson had refused Laffite's pirates?

As soon as Jackson entered New Orleans after the march from Mobile, he sensed that the locals didn't much appreciate the major battle planned for their home. He quickly decided to tread as lightly as possible on the citizens so that they wouldn't turn on his army and assist the approaching British. He considered imposing martial law as a security precaution to prevent spies from moving in and out of the city, but he knew that this could further antagonize the locals. The attitude of the people of New Orleans was a minor concern, however, compared to the larger problems facing Jackson.

There were rumors that the British had twelve thousand men, nearly six times as many soldiers as he had, and to make matters worse he was running low on materiel. Soon word spread of Jackson's predicament, and the general received two unusual offers of assistance. A group of free black men offered their services, but Jackson would not consider it. "Most of these men are most likely fugitives, and even those who are not would be incapable of serving as well as trained soldiers," Jackson said.

Then a more enticing offer materialized. Jean Laffite, a notorious pirate, was willing to barter the services of his men and his cache of weapons and supplies for the freedom of his jailed companions. Jackson had helped to capture several buccaneers while he had been in Mobile just weeks before—and as a former judge, he wasn't about to allow criminals to walk free out of jail under any circumstances. Jackson recorded later that he had considered arresting Laffite on the spot. "I could have taken LaFeet off to jail myself, and that is where he did belong," he wrote. Jackson didn't want his troops fighting against both the British and the privateers, however, and so he let the pirate slip back into the delta.

As American scouts and spies brought back word about the size of the British armada, Jackson began to realize that he had an impossible task before him. The British would be rested and well supplied. Many of them would be battle-hardened veterans of the Napoleonic wars. If twelve thousand men were on board the ships, the British might land a force of as many as ten thousand soldiers. Even if he could convince the local militia and volunteers to join him, Jackson doubted that he could find more than two thousand men, and all of those men would be useless if he couldn't find supplies and ammunition. Jackson knew that it would do the nation no good if he forfeited the lives of these loyal patriots in what would be, in the end, a lost cause. He then made a decision that he later said felt like stabbing himself with a knife: He marched his men out of New Orleans and back up the Natchez Trace.

Jackson arrived in Nashville, Tennessee, three weeks later and stopped at a tavern for the night. "How goes the war?" he asked the proprietress.

"Have you not heard, General?"

"What has happened?" Jackson could tell from the woman's manner that something important had transpired.

"General, the war ended weeks ago. The treaty was signed on Christmas Eve. We have no quarrel with the British any longer."

"My . . ." he said, unable to express what he was thinking. A weary Jackson sank into a chair, and he realized that fighting the British in New Orleans would not only have cost many men their lives, but that the battle would have been fought two weeks after the war had ended.

Jackson joined the thousands of others who were celebrating the end of the War of 1812. In subsequent weeks, however, Americans discovered that although the United States no longer had a quarrel with Great Britain, Britain had a quarrel with the U.S.

France had sold the Louisiana Territory to the United States in 1803 for $27 million in gold, which had gone to help a financially pressed Napoleon Bonaparte continue his war against Britain. But there was a problem with the deal: Spain's King Charles IV had reluctantly ceded the Louisiana Territory to Napoleon in 1800 in the Treaty of San Ildefonso, which had directed that France was never to give the land over to another nation. Napoleon sold the land to the United States just two years later.

Napoleon had also invaded Spain in his quest to conquer Europe, and Spain had joined Britain to fight the French. Eventually Napoleon was defeated and forced to flee France at the end of the Napoleonic War. Now that Jackson had chosen not to fight the British and Britain controlled the southern part of the Louisiana Territory, Britain declared that it intended to protect the interest of its ally Spain. Britain informed the United States that because the Louisiana Purchase had violated international law and was therefore an illegal act, Britain would keep the Louisiana Territory. In March 1815, Britain and Spain signed a treaty giving full ownership to Britain.

There were now two conflicting agreements on the ownership of the Louisiana Territory, a treaty between the United States and France giving ownership to America, and a treaty between Britain and Spain giving ownership to Britain. But the conflict was more than just dueling documents: Because the British had an army in New Orleans and a naval flotilla protecting the Gulf Coast, the United States could recapture the territory only by declaring war on Britain again.

In Nashville, Tennessee, Jackson began organizing a militia to return to New Orleans to fight the British, which he hoped would reverse the public opinion that he had lost the new territory. In Washington, war hawks such as John Calhoun began calling for reopening the war against Britain. But President James Madison and many in his cabinet were reluctant to fight again. The British had forced Madison to flee Washington for his life during the war, and this was while the best of the British military had been fighting in Europe. The possibility existed that a new war against Britain could cost the United States its independence.

James Monroe, who held the offices of secretary of state and secretary of war and thus was both the carrot and the stick personified, went to Britain to negotiate an end to the dispute over the ownership of the Louisiana Territory. In a complicated agreement, the United States ceded the Louisiana Territory to the British, and Spain transferred Florida to the United States. The unfortunate War of 1812 was finally over.

After the battle, the people of New Orleans were still smarting over the rough treatment they had received under General Jackson. When Jackson did not immediately repeal martial law after his victory—he was waiting for official word that the war was over—a local judge fined him one thousand dollars for violating a citizen's constitutional rights. Outside of Louisiana, however, Andrew Jackson was a hero. A snowstorm along the Eastern coast had disrupted mail deliveries to Washington and other cities, and word of Jackson's victory did not arrive until almost a month after the battle. The citizens had learned of the British landing in Louisiana, but then there was no news for weeks. When word of the victory did arrive, Jackson became an instant American hero. The citizens of the states could hardly believe what had happened. The British Royal Navy was the most powerful in the world, and the armada was claimed by some to be the largest that had ever sailed up to that time. The British had attacked with what was considered a huge force of twelve thousand men (small compared to the European armies of Napoleon and Wellington, but enormous for an army in North America). As word reached the cities, the citizens poured into the streets to celebrate, and it was said that the torches of the impromptu parades turned night into day in the streets.

A proclamation issued by the governor of the Louisiana Territory praised Jackson's "undaunted courage and patriotism," and noted, "Natives of different states, acting together for the first time in this camp, differing in habits and in language, instead of viewing in these circumstances the germ of distrust and division . . . have reaped the fruits of an honorable union."

Napoleon most likely did violate the Treaty of San Ildefonso in the Louisiana Purchase. But because Andrew Jackson was able to achieve an incredible defeat of the world's superpower at New Orleans, the rights to the vast territory were secured.

Chapter 8

The Election of 1824

❦

In 1824 the nation elected a vice president but failed to supply enough votes to select a president, and congressional deadlock ensued. Eventually it would fall to one congressman to cast the deciding vote. But what if the deadlock had continued?

IN junior high civics class, every American learns that one vote could make a difference in an election. That lesson, along with most other lessons of middle school, is often promptly forgotten. But in the election of 1824, the election really did come down to just one vote, that of Congressman Stephen Van Rensselaer of New York.

That election was one of just two decided by the U.S. House of Representatives. The election occurred during what historians label the Era of Good Feelings, so called because there was only one true political party, but it would be difficult to come up with a more misleading title for the times. In the election of 1824, there may have been just one political party, Thomas Jefferson's Democratic-Republicans, but there were plenty of presidential candidates to choose from: Five men made a run at the office. From New England there was John Quincy Adams, the son of our nation's second president,

John Adams; representing the South were John C. Calhoun of South Carolina and William Crawford of Georgia; and from the frontier western states were Henry Clay of Kentucky and Andrew Jackson of Tennessee.

Early in the campaign, Calhoun realized that he stood the poorest chance of becoming president and decided to set his sights on the vice presidency instead. Calhoun was a master politician full of ambition. Just thirty-nine years old, he thought that the vice presidency would be a good place from which to launch an eventual run for the presidency. To improve his chances, Calhoun tried a political gambit unparalleled in American history: He ran for vice president on two tickets.

Calhoun had approached the front runner in the contest, John Quincy Adams, and offered to serve as his vice president. With this arrangement agreed upon, Calhoun then went to Andrew Jackson, the other leading candidate, and offered to run as *his* vice president. Jackson was said to be amused at Calhoun's ingenuity, but Adams was concerned. "Calhoun's game now is to unite Jackson's supporters and mine on him for vice president," Adams wrote to one of his supporters. "Look out for breakers."

The leading candidate was William Crawford, secretary of the treasury. Unfortunately, a year before the fall election, the fifty-one-year-old Crawford suffered a severe stroke and was unable to campaign. Each of the four remaining candidates, Calhoun included, had reason to believe that he would be elected president.

Jackson was also well known nationally because of his heroics at the Battle of New Orleans in the War of 1812, but many people were put off by Jackson's rough-hewn manner and his controversial personal life (which included shootings and alleged bigamy by his wife). John Quincy Adams, on the other hand, was a straightlaced intellectual. He held the view, not shared outside his native New England, that he was the natural successor to the current president, James Monroe. On the opposite end of the morality scale from Adams was Henry Clay, who had a reputation as a frequent gambler. (When one prudish woman tittered her sympathies to Mrs. Clay for Henry's gambling, Mrs. Clay replied, "Oh, I don't know. He always wins.") Clay was also known as one of the most eloquent congressional orators ever—one author said that his speeches were as "musical as Apollo's lute"—who by age forty had become the Speaker of the House.

The 1824 vote split somewhat along regional lines. Jackson was popular in several Southern states, the Midwest, and the mid-Atlantic region, Adams held the New England states and New York, Crawford won the Southeast, and Henry Clay came in a close fourth, winning the states of Kentucky, Missouri, and Ohio.

When the election results trickled into Washington in mid-December, they showed that Jackson had won the largest number of votes and won the most electoral college votes, but that he failed to win a majority of either. To be precise, Jackson received 152,901 votes for 99 electors; Adams received 114,023 votes for 84 electors; Crawford received 46,979 votes for 41 electors; Clay received 47,136 votes for 37 electors. For the first time ever, a U.S. presidential election would be decided by the House of Representatives, which would pick among the top three candidates.

Calhoun's political maneuvering had succeeded. By running on the ballot with both Adams and Jackson, Calhoun received two-thirds of the electoral votes for vice president, enough to secure his election.

There was strong sentiment to give the election to Jackson. Thomas Hart Benton, the great legislator from Missouri, summed up the feelings of many in the country when he said that because Jackson had won the most votes, Congress should be guided by "the demos kraeto principle" and make Jackson's election official. However, to the dismay of the electorate, it soon appeared as though the politicians were busy cutting deals instead of weighing what would be best for the country. One Missouri congressman, for example, had a brother who was a judge who had recently killed another judge in a duel. The congressman arranged to give his vote, the only electoral vote for Missouri, to Adams, even though almost no one in the state had voted for Adams. In exchange, the congressman's brother was to be renominated to the bench if Adams became president.

Because Clay would not be in the running, and his thirty-seven electoral votes were up for grabs, he suddenly became the most popular man in Washington. Clay was amused at his newfound celebrity, saying that the various candidates "all believe that my friends have the power of deciding the question, and that I have the power of controlling my friends." When Clay attended a dinner and found Jackson and Adams sitting with an empty chair between them, Clay loudly joked, "Well, gentlemen, since you are both near the chair, but neither can occupy it, I will slip in between and take it myself."

For his part, Clay summed up the sentiments of much of the Washington establishment when he voiced his dismay at Jackson's appeal: "I can't believe that killing twenty-five hundred Englishmen at New Orleans qualifies him for the various, difficult, and complicated duties of the Chief Magistry." Clay was worried about what Jackson might do to the country, and more personally, he was concerned about what a Jackson presidency might do to his power base in the western states. Clay soon announced publicly his support for Adams, and his decision was immediately condemned. The Kentucky state legislature voted to support Jackson, but Clay said they had no right to direct him or any of the Kentucky electors in their votes, and he convinced his fellow Kentucky congressmen to vote for Adams despite the fact that Adams had not received a single vote in the entire state.

Clay's support for Adams soon became a national scandal. People suspected the Kentucky congressmen were voting more for personal gain than in the best interest of the country. Rumors soon began circulating that Clay had given his support to Adams in exchange for the position of secretary of state, a charge made all the more credible by Clay's boast that if he wanted a place in the Adams administration, "I can enter into it in *any* situation that I may please." Because this arrangement appeared to be a bald political move for personal gain at the expense of the people who had voted against Adams, the situation was soon labeled "the corrupt bargain."

Clay met with Adams a second time and told him of the outrage building in the country. "This morning I received an anonymous letter . . . threatening civil war if Jackson is not chosen," he reported.

The vote in the House was as close as imaginable. The election came down to the vote of the thirty-four representatives from New York. Seventeen of them announced support for Adams, and sixteen announced support for Jackson or Crawford. One person, Stephen Van Rensselaer, was undecided.

"I am too old to engage in any active electioneering . . . I am exhausted and am now disposed to glide with the stream," the sixty-year-old Van Rensselaer had said before the election. Early in the campaign he had supported Henry Clay; then he said that he favored John Calhoun. After the election was thrown into the House, Van Rensselaer confessed, "I feel inclined for Old Hickory." But after a meeting with Adams, Van Rensselaer promised John Quincy his support in the vote in the House. Crawford's campaign manager, Martin Van

Buren, then put his considerable political skills of persuasion to work on the old man, and Van Rensselaer changed his mind and announced that he would vote for Crawford.

Early in the morning of the vote in the House, crowds of people braved a snowstorm to attend the historic event. Van Buren met with Van Rensselaer again that day, and Van Rensselaer informed Van Buren that he intended to vote for Jackson, but again Van Buren convinced him to support Crawford, and as he went to the Capitol, Van Rensselaer promised to uphold his pledge to support the Georgian. But he was in tears trying to decide among the candidates: "The election turns on my vote—one vote will give Adams the majority—this is a responsibility that I am unable to bear. What shall I do?"

A few days later Van Rensselaer offered an explanation of how he came to his decision. After Speaker Clay had called the roll and summoned the representatives to vote, Van Rensselaer said, he had laid his head on his desk in prayer as the ballot box made its way around the House chamber. As he raised his head, he saw a discarded ballot that someone had dropped on the floor. The ballot was for John Quincy Adams, and taking this as a sign from above, Van Rensselaer voted accordingly. Adams was elected president on the first vote.

WHAT IF?

What if Van Rensselaer and his brethren in the House of Representatives had not been able to reach a decision on the first ballot?

Summoned to vote, Van Rensselaer gave in to Martin Van Buren's pressure and decided for Crawford. Van Buren knew this meant that all of the deals and agreements made by the men in Congress were now for all practical purposes invalid. A new vote count meant that the deal making could start from the beginning. Van Buren decided to pay a visit to Mr. Adams.

"Mr. Adams, or should I assume the inevitable and address you as Mr. President?" Van Buren said as he entered Adams's library. Van Buren told Adams that he had stopped by to see if there was any "responsible manner for gentlemen to remove ourselves from this difficult situation." Adams was wary of Van Buren and his tricks, but he listened carefully when Van Buren began to speak about Henry Clay.

"The election will be yours tomorrow," Van Buren said. "Why, already Mr. Clay has lined up most of the necessary votes. He has assured the other representatives that your . . . what was the term he used? The presidential partnership—that was his term, partnership—of Adams and Clay will be most receptive to their concerns." Van Buren could tell by the fire burning in Adams's eyes that his volley had found its target. Playing on Adams's insecurities, Van Buren began letting it slip that Clay was telling people that he—Clay!—would be the real power in the new administration.

Van Buren then carefully packaged Adams's responses—that Clay lacked the diplomatic skills necessary for secretary of state and that, furthermore, Adams couldn't think of any other place where Clay would help his administration, either—and passed them along to Clay's friends. Soon, Clay and Adams confronted each other, and the two prideful men parted angrily without resolving their differences.

Clay would no longer support Adams, but he wasn't about to support his rival for control in the western states, Jackson. Instead he began urging his supporters to give their votes to Crawford. The old Georgian was a more palatable choice to the voters of Kentucky and Missouri than Adams anyway, and caught up in the spirit of a mini-revolt, they eagerly agreed. Van Buren's Machiavellian maneuvers had worked, to a degree. He had succeeded in creating the logjam that he wanted, because now the three candidates all had nearly equal numbers of electoral votes pledged to them.

The inauguration was to be held on March 4, just three weeks after the first vote in the House. At first Speaker Clay and the other congressmen had not been worried about the date. They scheduled a second vote for February 10, and another for February 11, and so on until by mid-February, worry was beginning to set in.

Van Buren's logjam—designed to provide time for him to get enough electors to switch to Crawford—would not succeed, because he was not the only person who had an interest in stealing the election. Watching quietly from the sidelines, John C. Calhoun soon inserted himself into the deliberations.

Just a week before the inauguration, Calhoun had orchestrated an invitation to address the members of the House in an effort to break the electoral lock that gripped the chamber. Using his dashing manner, good looks, and passionate oratory skills to their fullest advantage, Calhoun stood at the Speaker's post and reminded the congressmen of the danger to the nation if they did not pick a president by March 4. There was only one way to settle the dispute and offer the American public a president that they would have full confidence in: The members of Congress must select him, John Calhoun, as president.

There was an immediate roar of disapproval in the House as many of the most strident supporters of Jackson,

Adams, and Crawford shouted their objections. But as Henry Clay, himself visibly shaken by Calhoun's words, pounded his gavel for order, a murmuring arose from the seats and the galleries like the hum of locusts in a field. Calhoun's proposal was damnably self-serving, but maybe he was right. Two-thirds of the country had voted for him for the nation's second office, where he would be ready to step up to the presidency if that office became vacant. It certainly appeared that there was a good chance that the office of the presidency would be vacant on March 5. It wasn't the normal circumstance under which a vice president was elevated to the presidency, but perhaps this was the best solution. After two more hastily arranged votes, Calhoun was elected president of the United States.

The result was disastrous for the country. Although Calhoun did swear to uphold the Constitution, in his reading of the document, each state was empowered to leave the Union at any time, a process known as nullification. It wasn't long into Calhoun's term, in 1826, that four states, South Carolina, Georgia, Alabama, and Florida, fearing that the United States might soon prohibit slavery, did nullify their association with the Union.

Calhoun insisted that the secession of the states was legal and refused to send troops to quell the rebellion, overruling the strident voices in Congress who were insisting that he mobilize the Army. The Calhoun presidency allowed a split in the nation that became known as "the Great Divide."

The election of 1824 may have been resolved quickly, but it wasn't resolved without rancor. The uproar over the election results lasted for months afterward. Congressman John Randolph of Virginia said that the union between the straight-laced Adams and the gambling Clay was a "combination, unheard of till then, of the puritan with the blackleg." Clay and Randolph actually fought a duel over the comment, but both men survived. Clay would run for president two more times, but he was stained with the perception that his vote had been bought and sold in the 1824 election, and he never achieved the office that he desired so much.

Adams didn't fare much better. He won the presidency but found himself in a deadlock with Congress and subjected to abuse outside the halls of government. When Adams and his wife attended a stage play shortly before the inauguration, the actors on stage suddenly began making jokes about the election and then began singing a song about Jackson's victory in New Orleans, which brought cheers from the audience. According to someone who attended the play with Adams, the president-elect suffered the ridicule "with death-like silence."

Andrew Jackson, who was able to play the role of political martyr, found that he was more popular in defeat than he had ever been in victory. In 1825, the Tennessee legislature took the unprecedented step of immediately nominating Jackson for the 1828 election. Jackson ran against Adams in 1828 and won in a landslide.

Crawford's advisor, Van Buren, was amused that Clay and Adams seemed surprised by the animus that poured out of the electorate over the election outcome, and he was further amused at the naïveté of the citizens who were outraged over the political maneuvering. Van Buren explained, "Take my word for it, there is no honorable profession extant, in which there are more knaves enjoyed than in this same profession of politics."

Chapter 9

Messages Sent by Lightning

❧

Acclaimed artist Samuel Morse decided to give up his painting to pursue work on a communication device that he claimed used lightning to send messages. What if he had stuck with painting?

IN the spring of 1825, Samuel Morse received a tragically delayed message.

Morse was a struggling artist living in New York, sleeping in his studio to save money, while his wife Lucretia and his four children waited for him in New Haven, Connecticut. That spring, Morse was chosen to do a portrait of the Marquis de Lafayette, the famous Revolutionary War hero. For Morse, the Lafayette portrait meant that he would finally have enough money to bring his family to New York. After spending a few days at home in New Haven, a trip that allowed him to see his four-day-old baby, Morse left for Washington, D.C., where he was to meet with Lafayette.

As soon as he arrived in Washington, Morse received a letter from a family friend in New Haven telling him that his wife was suffering from complications from her recent delivery. "Your dear wife is convalescent," the letter said. "We shan't hurry her from her chamber at this season—the children all continue as hearty & playful as when you left them . . ." After reading this news, Morse began preparing to quickly finish his painting of Lafayette and hurry home to be with Lucretia.

Before he left Washington, Morse attended a party at the White House held by President James Monroe to honor the resolution of the disputed election of 1824 between John Q. Adams and Andrew Jackson. Morse watched Jackson shake Adams's hand, and he was eager to tell his wife about his witnessing the important event. But before Morse was able to complete his business in Washington, he received a second letter telling him that his wife had died. In fact, she had died the day after the family friend had posted his first letter to Morse. Morse immediately dropped everything and rushed to New Haven, but by the time he arrived there he discovered that the funeral had taken place several days earlier.

Such delays in communication were commonplace in Morse's time. In the early 1800s, information rarely traveled any faster than it had in the time of the Roman Empire. Communication and transportation were inseparable—information didn't travel any faster than a person on a horse or ship could deliver it. There were attempts to speed up the process by using carrier pigeons, cannon signals, signal flags, and other technologies, but these methods were unreliable and cumbersome and certainly not available to the general public. In the early 1800s, railroad lines sped up the transfer of information somewhat—some enterprising newspapers put typesetting equipment on railroad cars so that newspaper stories could be typeset as the news traveled to the newspaper's office—but by and large, information traveled at a trot.

After his wife's death, Morse spent several years searching for his role in life. In addition to painting, he tried politics for a while. Morse returned from a trip to Italy with a bigoted view of Catholicism, and he latched onto a nativist, anti-Catholic, anti-immigrant political view that was popular in the 1840s. He wrote and published a few bigoted pamphlets, and in 1836 he made a run for mayor of New York, finishing fourth in a four-way race.

Returning from a trip to France in 1832, Morse began discussing electricity with a scientist, Thomas Jackson, who was also making the voyage. Morse began peppering Jackson with questions about recent developments in the field, because while on board the ship, Morse had conceived of a way to transmit information using electricity. Unlike most inventors who tinker with a device for years before it finally works, Morse had a vision of the entire contraption, and when he arrived back in the United States he decided to give up his career as a painter and devote his time to creating this communication machine.

Other than scientists, almost no one in the country understood electricity. It was generally thought of as something Ben Franklin had experimented with using a kite and a key. When Morse told his friends about what he was working on, many thought that his idea was crazy. James Fennimore Cooper, after hearing of Morse's plan, told his family that it was outlandish behavior for an artist even to think that he could send messages on the wings of lightning.

Morse wasn't alone in his attempts to use electricity to send messages. In fact, more than sixty devices had been built before Morse invented his telegraph. The best of these was created by the Englishmen Charles Wheatstone and William Fothergill Cooke, who had developed a telegraph system that used magnetic needles to point at letters on a board. Their system was much slower and more prone to errors than Morse's electromagnetic system, however.

Morse's invention consisted of three elements. There was the sending device, which transmitted signals by opening and closing an electric circuit. The second part was a receiver that recorded the electric pulses as either dots or longer dashes. But perhaps the most important part of the invention was the Morse code, or more properly the Morse alphabet, which converted all numbers and letters into a binary system of dots and dashes. The Morse alphabet was the software, or operating code, that the other inventors had been missing.

Morse thought that his device would primarily be used by government, and he went to Congress to ask for funds to continue the device's development. When Morse demonstrated his invention, one congressman from Tennessee was so sure that it wouldn't work that he told Morse to send the message "[President] Tyler deserves to be hanged," a message that Morse was able to transmit. Despite this success, a senator from Indiana said that he carefully studied Morse's face to determine if the inventor was insane.

When the funding for Morse's invention came before the conference committee, Morse watched from the gallery as congressmen debated how to fashion the final funding bill. Cave Johnson, another congressman from Tennessee, said that since they were going to fund magnetism, Congress should also fund mesmerism, a belief that ailments could be cured through hypnotism. A congressman from Alabama joined in the teasing. If they were going to fund magnetism and mesmerism, he said, Congress should also support Millerism, a religious belief that the Second Coming of Christ would occur in 1844. The congressmen were all enjoying a good laugh at their wit, while Morse sat in the gallery with his hand on his forehead. "I have an awful headache," he said.

Despite the obvious facetiousness of a proposed amendment funding research on mesmerism and Millerism, twenty-two apparently confused congressmen voted in favor of funding all three.

Nonetheless, the amendment was defeated, and it appeared that the funding for a telegraph line for Morse's device would suffer the same fate. A rival painter wrote to a friend that Morse "has been all winter at Washington trying hard to push his 'dunder & blixen' Telegraph thru Congres . . . I am afraid that Uncle Sam will be found lightning-proof in this case. . . ." Then, surprisingly, a supporter of Morse's was able to get the bill to the floor of the House for a full vote. When the bill was announced, 70 of the 242 representatives left their seats so that they wouldn't have to vote for a device that they didn't understand. The others didn't let their ignorance interrupt them, however, and the bill to fund the world's first telegraph line was voted through, eighty-nine to eighty-three. It was not a popular vote with many constituents, however. Congressman Lew Wallace, later a Civil War general and author of *Ben-Hur*, lost his House seat in the very next election because his Indiana constituents were upset that he had voted to spend thirty thousand dollars of public money on this ridiculous machine.

The money was spent to construct a telegraph line from Baltimore to Washington, and the enthusiasm for the telegraph spread almost as fast as the messages themselves. Within eighteen months there were nearly 2,000 miles of telegraph wire strung across the country, 10,000 miles of wire within five years, 23,000 miles within eight years. Not only was the telegraph amazingly fast from point to point, it could send messages to many points simultaneously, which was a boon to many businesses. "It seemed like the nervous system of the nation," a traveler on the Mississippi River marveled in 1850, "conveying, quick as thought, the least sensation from extremity to head . . .

and by these wires stretched across the Mississippi, I could hear the sharp, quick beating of the great heart of New York."

WHAT IF?

What if the painter Samuel Morse had decided to abandon his vision for the telegraph?

1832. Morse was in his studio early, hoping to catch a bit of inspiration in the morning quiet. Nothing but quiet until the clock in his studio sounded, and he was reminded with each tick that no ideas spurred his enthusiasm enough to move him to put brush to canvas.

In the week since he had returned from his trip to France, every day in his studio had been spent staring at the half-finished canvases. The conversation with the scientist Thomas Jackson about the possibilities of electromagnetic devices had been stimulating, but once back home the harsher realities of his world faced him. He had four children to provide for, and he was just beginning to make his reputation as an artist. This wasn't the time to be chasing electron dreams.

Still, it was easy for his friend James Cooper to scoff—Cooper had already made a name for himself internationally as a novelist. Morse belonged to the same social club as Cooper, the Bread and Cheese Club, but there was a gulf between their reputations. Cooper could laugh the easy laugh of a successful man, but Morse's own career had a bitter flavor. He had been a failure at his forays into politics, and he was reluctant to begin chasing a new career as a scientist here at midlife. What else was there? Art was the thing that he had proven to himself and others that he had a talent for, and so it was in this field that Morse decided that he must intensify his focus and redouble his efforts.

For a few years Morse did concentrate his energies on his portraiture, only pausing to rethink his decision about the electromagnetic telegraph when he would occasionally come across his faded notes that he had scribbled out while on board the Scully years before. But Morse had an insatiable mind, and nearly a decade after his brush with electricity, another new technology grabbed his enthusiasm, one more clearly in line with his artistic profession. In the early 1840s Morse began studying the new techniques of photography. He began spending more and more time with the plates and chemicals, and by the mid-1850s, he was well-known across America as a photographer of famous people and places.

The problems of communication that had such a profound effect on Morse earlier in his life had been somewhat eased by the same decade. The railroad, which now stretched through most of the Northern states as far west as Chicago and included a line to New Orleans, allowed a message to travel from New Orleans to New York in just days, instead of the weeks it had taken previously.

The increased speed of communication was a boon to the young nation as the population spread into the Great Plains territories on the other side of the Mississippi River. In 1848 California became part of the United States, and after the discovery of gold in the territory the same year thousands of people began moving from the Eastern states to the new, rich territory.

At the beginning of 1849, San Francisco was a small frontier city of two thousand people, but by December it was a city of nearly fifty thousand people, rivaling, in numbers, some of the largest cities in the East. The forty-niners who had chased the dreams of wealth to California were men who were remaking themselves in a new land, and they had little patience for the ways of the older nation back East. They were mostly young, many of them impulsive and more than a few hotheaded, and all of them flush with either money or dreams of it. The entrepreneurs of information in San Francisco saw themselves as a new type of people. They dressed more casually than the businessmen, and they had contempt for the established ways of conducting business back East, with its conservatism and its serious-minded emphasis on the buffoonery of Washington politics. The young men, many of whom had become fantastically rich before they reached their thirtieth birthdays, had a deadly arrogance bred of wealth and youth.

California became a state in 1850, but a sentiment soon began to grow that California was a different type of state than the others, one whose problems and opportunities were unlike those of the states on the Atlantic Coast. During the 1860s, as the priorities of the United States became eclipsed by the Civil War, more people came to California, many of whom, like the journalist Samuel Clemens, were looking to avoid the conflict that was dividing the states back East. As the entire nation east of the Rocky Mountains dealt with the war and its after-shocks, word of the unfolding tragedy came slowly to the people in California. News of important battles and decisions arrived sometimes two months after the events back East, increasing the feelings of isolation felt by the people on the Pacific Coast.

By the mid-1860s, many people in California began

thinking that the state deserved its own government, one that was more concerned and responsive to the new state, and a separatist political party, the Golden Bear Party, began attracting more and more people to its political rallies.

The leaders of Congress and President Ulysses Grant heard and read sporadic reports about this small group of troublemakers, but they were caught completely off-guard when Cornelius Feingold, a millionaire businessman and leader of the Golden Bear Party, was elected governor. Feingold almost immediately had the military posts in the state seized, and declared California's independence.

Loyalists immediately set sail out of San Francisco Bay to race to Washington to inform President Grant of the rebellion. They stopped in Central America to cross the Isthmus of Panama, then sailed to New Orleans, where they boarded a train bound for Washington. The news that the government of California had declared its independence was finally delivered to President Grant two months after the fact.

In the years following the Civil War, the federal army had dwindled to just a few thousand men, most of whom were scattered at small forts and outposts throughout the plains and desert southwest. Even by this time the old general knew that the rebels in California would have sealed off the few passes through the mountains. What's more, there was little sympathy in the Eastern states to fight another war to force a wayward state to remain in the Union, and there was the real possibility that frustrated Confederates would race to California to join in a second rebellion as fast as any federal troops could arrive to quell the insurrection. Forcing California to remain in the

Union was hopeless, and in 1871 Grant agreed to recognize the new nation of California.

The inability to hold California in the Union confirmed what James Madison had written in Federalist Paper No. 10: "Extend the sphere and you take in a greater variety of parties and interests; you make it less probable that a majority of the whole will have a common motive . . . or, if such a common motive exists, it will be more difficult for all who feel it to discover their own strength and to act in unison with each other." It was simply impossible to hold a nation together across an entire continent when there was no way for the people to communicate.

Of course, Cornelius Feingold never existed, nor did California's secession. But it is possible that similar events would have occurred if it weren't for the telegraph. Morse's invention had not only had as great an effect on society as the invention of the printing press, it was a technology that allowed a nation the size of the United States to exist peacefully. By 1861, telegraph lines in the United States reached from coast to coast, eight years before a transcontinental rail line would complete the connection of east to west to much greater fanfare.

Morse, of course, became famous and wealthy from his invention. He actually was one of the first Americans to embrace photography, and he developed techniques that allowed photographs to be taken quickly enough to photograph people. Among the students in Morse's photography studio was a young Matthew Brady, who would become a famous photographer of the Civil War. But it was the telegraph that made Morse a name recognized around the world.

Morse's invention had one more direct effect on the United States: It led to the purchase of Alaska. William Seward, secretary of state under Ulysses Grant, thought that it would be easier to run telegraph lines up the Pacific Coast, across Alaska, the Bering Strait, and Russia, and into Europe than it would be to try to place a telegraph wire under the Atlantic Ocean. Seward was wrong, because while negotiations for the purchase of Alaska were beginning, a telegraph line was strung across the Atlantic Ocean. But by this time Russia and the United States were so set on the idea of transferring Alaska that it was a fait accompli. Thanks to the telegraph, the young nation of the United States grew quickly to become one of the largest countries in the world.

Chapter 10

Conquering Mexico

⊷

After going to war with Mexico, President James K. Polk was presented with an illegal treaty, which he accepted. What if he had demanded more?

IN November 1847, Nicholas Trist learned that he had been fired.

Trist, a clerk in the State Department, had been sent to Mexico by President James K. Polk to negotiate an end to the Mexican-American War. He had been chosen because the president wanted to keep the mission a secret (sending the secretary of state or another high official would have alerted the press), and because the peace deal had been worked out. All Trist needed to do was get the signatures of the Mexican authorities on the treaty. However, the situation in Mexico changed when the Mexican army, led by Santa Anna, refused to stop fighting. It quickly became obvious that more experienced diplomatic skills were needed, and Polk decided to recall Trist.

But Nicholas Trist refused to leave Mexico. He thought that he was making progress toward peace, and because he had always believed that he would one day make history—Trist had married Thomas Jefferson's granddaughter, and he behaved as if it were he, and not his wife, who was descended from the famous statesman—Trist decided that this was his moment to achieve greatness. He wrote a harsh letter to Polk implying that the president didn't want peace and that it was up to Trist to stop the war. Trist's missive was "arrogant, imprudent, and very insulting to the government, and even personally offensive to the President," Polk wrote in his diary.

James K. Polk of Tennessee had been a Speaker of the House, but people were surprised when he was nominated for the presidency in 1844. "Jem Polk being President of the United States!" chuckled a Tennessee congressman. "We are more disposed to laugh at [the idea] here than to treat it seriously." The dark horse Polk surprised many by defeating Martin Van Buren in the election that fall.

Polk, a political descendant of Andrew Jackson, was sometimes known as "Young Hickory," but he lacked Jackson's flair. He was a short, dour man who banned both alcohol and dancing in the White House, and he was known to regularly put in eighteen-hour days in his office while president. Sam Houston said that the main problem with Polk was that he drank too much water.

Polk had run for president promising to do five things: First, he said that he would serve only one term. He also promised to settle the dispute over the Oregon territory, secure California for the United States, reduce tariffs, and reestablish an independent treasury.

Polk was able to accomplish all of his goals (making him a true historical curiosity), but it was the acquisition of California and the southwestern United States that earned him a controversial place in American history.

The dispute over the southwestern boundary of the United States was an old one. President Thomas Jefferson had claimed that the Rio Grande was the proper border, although Mexico had claimed it lay much farther east and north; President John Quincy Adams had authorized ambassador Joel Poinsett to offer Mexico $1 million to agree upon the Rio Grande as the border. Poinsett returned to Washington with a beautiful Mexican plant that was named for him, but with no treaty. Relations between the United States and Mexico remained strained.

Although the region from Texas to California was claimed by Mexico, in reality it was largely unsettled land that had about an equal number of American and Mexican settlers. When Texas won its independence from Mexico in 1836, and shortly thereafter asked to be included in the United States, an armed conflict over the territory and the southwestern border was almost inevitable.

Mexico did little to ease the dispute. Just two days after Polk was inaugurated in March 1845, Mexico withdrew its ambassador to and broke off official relations with the United States over the issue of Texas. It was a foolish move, because Texas had long since gained its independence and Mexico had no chance of regaining the territory, although Mexico had not yet officially recognized Texas's independence.

Mexico and the United States had other conflicts. Mexico had borrowed large sums of money from U.S. citizens and then refused to repay the loans. Also, in the frequent civil wars in Mexico, American property was often confiscated or destroyed without compensation. Polk offered to forgive all Mexican debts in the United States if the country would recognize the Rio Grande as the rightful border between the two countries, but the Mexican government refused to talk with the delegate

Polk sent to hammer out the deal. Mexico continued to insist that the border was the Nueces River, one hundred miles to the northeast of the Rio Grande. With diplomacy going nowhere, just over a year into his administration Polk ordered the sixty-year-old general Zachary Taylor to move American troops to the edge of the Rio Grande. It was a provocative action most likely designed to provoke a hostile response.

The Mexican government immediately obliged. The same day that the U.S. troops arrived—April 25, 1846—the Mexican army attacked a small U.S. unit and killed eleven soldiers. Word of the attack reached Polk on May 9, and two days later Polk declared to Congress that since the Mexicans had attacked the United States, a state of war already existed between the two countries and that he was going to respond.

John O'Sullivan, editor of the *Democratic Republic*, said that whatever the means, America was preordained to own the land. "Yes, more, more, more!. . . . Till our national destiny is fulfilled and the whole boundless continent is ours." This national destiny—soon polished with the patina of Providence and called Manifest Destiny—became a national rationalization for the war. "It is for the interest of mankind," wrote then-journalist Walt Whitman, "that [the United States'] power and territory should be extended. The farther the better."

Many politicians opposed Polk's war, but not all were willing to say so. The Whigs remembered how the great Federalist Party of Hamilton, Washington, and Adams had simply disappeared after it had opposed the War of 1812. The Whigs quickly learned that it is political self-immolation to try to stop Americans ready for a fight, and most Whigs kept their personal objections hidden. Speaking of his newfound support for the war, one Whig congressman joked bitterly that he was now also willing to express his support for "pestilence and famine."

Not all Whigs kept quiet about about what they viewed as a provocation designed to allow the United States to steal land from Mexico. A freshman congressman from Illinois named Abraham Lincoln introduced legislation into the House of Representatives that demanded to know the exact spot where Mexicans had attacked Americans to start the war. Lincoln's legislation went nowhere, as did Lincoln's attempt for reelection in the next election.

Lincoln was nearly alone in his opposition: the declaration of war passed the House 174 to 14 and the Senate 40 to 2.

In hindsight, Mexico's eagerness to go to war with the United States was ill-advised. Mexico had already lost Texas, and it had only a tenuous hold on California. But at the time Mexican leaders assumed they would win a quick victory, and they weren't the only ones who thought so. Mexico had a standing army four times larger than that of the United States, and in the words of the *Times* of London, their armies were "superior to those of the United States." The mountainous land, the vast, hot deserts, and the Americans' unfamiliarity with the terrain would be an obvious advantage for the Mexican army. Many European observers expected Mexico to emerge the victor.

Zachary Taylor quickly won several battles against the Mexican army and was able to move south as far as Monterey. He immediately became a hero in the United States as his successes were reported, and people soon began suggesting him as a potential presidential candidate in the next election. President Polk became jealous of Taylor's popularity and frustrated at the fact that Taylor wasn't advancing farther south to gain more territory, so Polk split Taylor's army and gave half of his men to Winfield Scott.

The Mexican general Santa Anna soon tried to take advantage of Taylor's weakened army and marched toward it. Santa Anna arrived at Buena Vista with twenty thousand men and appeared to have Zachary Taylor cornered and the war won. Taylor, however, wasn't ready to quit. When he received Santa Anna's demand for surrender, Taylor replied, "Tell him to go to hell." Taylor's Kentucky troops rallied, and as the American artillery pounded the Mexican troops, Santa Anna was forced to retreat. Taylor soon began an overland march to Mexico City across desert so hot that his men marched in their underwear.

Meanwhile, in the northern territories claimed by Mexico, another force of men left Fort Leavenworth, Kansas, and marched toward Santa Fe. The Mexican governor of New Mexico declared that "it is better to be thought brave than to be so" and fled the city, allowing the Americans to capture the New Mexico territory without firing a shot. In California, John C. Frémont took his orders to "watch over the interests of the United States" quite broadly. Frémont assembled a small army, captured the town of Sonoma, and declared California an independent state, with the expectation that it would be absorbed into the United States once the war was over.

Winfield Scott had taken his half of Taylor's divided army and sailed to the port of Veracruz. Scott and his eleven thousand men easily captured the city and began to march northwest to Mexico City, a journey that was much shorter than the route that Taylor's men were taking. As they approached the city, Scott found that Santa Anna had placed twenty-five thousand men in positions above the main pass into the city, and he realized that he faced certain defeat if he tried to force his way through. Fortunately for the Americans, a young captain named Robert E. Lee was able to find a trail around the pass that led to a peak overlooking the Mexican army, and Scott was able to move there and destroy Santa Anna's army by raining artillery shells down on them. Scott quickly moved into Mexico City, where American troops stomped into the National Palace, the famed "halls of Montezuma" of the Marine Hymn.

In September 1847, the war was over except for the paperwork.

༄

The treaty negotiated and signed by Nicholas Trist received immediate criticism once it arrived in Washington six months after the fall of Mexico City. Polk was not pleased that Trist had disobeyed his orders and had continued to negotiate with Mexico, and many in Washington insisted that the treaty be ignored.

There was strong opposition to the treaty by abolitionists in the North, led by Daniel Webster, who feared that Southerners would expand slavery through the entirety of Mexico. But there was equal opposition from another powerful faction that opposed the treaty because they thought that the United States should take over more of Mexico. Jefferson Davis wanted to annex everything north of Mexico City. Sam Houston wanted to capture everything north of Veracruz. Secretary of State James Buchanan (who would become president) and Secretary of the Treasury Robert Walker broke with their administration and spoke out against the treaty because they thought that it didn't annex enough territory.

There was also a paternalistic theory that a takeover by the United States would be for the Mexicans' own good. The Mexican government was one of the least stable in the world. From the moment of their independence from Spain in 1821 until the start of the war with the United States—which the

Mexicans called *la invasión norteamericana*—every elected president except one had been overthrown. In the three years of the war the situation hadn't improved: Jose Joaquin Herrera was overthrown by Maiano Paredes y Arrillaga, who was overthrown by Jose Maiano Salas, who was replaced by General Santa Anna, who resigned after the fall of Mexico City and was replaced by Manuel Peña y Peña. A takeover by the United States could bring stability to the region.

Polk, however, finally decided that the treaty gave the United States enough, and he fought for its passage in the Congress.

In Mexico there was also a debate over the ratification of the treaty, which gave away half of the country's claimed territory. Those in favor of the treaty pointed out that the land in question was worthless and virtually uninhabitable, and, what's more, a failure to accept the treaty could mean a renewed war in which the United States would take over all of Mexico. Because the treaty allowed the Mexicans to keep their country, the Mexican negotiators wrote, "it may be more properly called a treaty of recovery rather than one of alienation."

It was a flawed treaty. In 1853 the United States had to purchase a portion of southern New Mexico (in the Gadsden Purchase) to settle a dispute caused by the treaty, and in the 1914 the U.S. Navy briefly occupied the port of Veracruz when the conflict threatened to boil over. Congress has almost routinely been called upon to settle disputes arising from the treaty, the most recent incident occurring in 1998.

The war had cost the United States thirteen thousand men to battle and disease and, less significantly, nearly $100 million. For this the United States gained a million square miles (including Texas), doubling its area and cutting the nation of Mexico in half. The land would eventually be carved into the states of California, Nevada, New Mexico, Utah, Arizona, Texas, Colorado, and Wyoming.

WHAT IF?
What if James K. Polk had rejected Trist's illegal treaty?

"Damn him. Damn him!" Polk slammed his fist down on the desk hard enough that Secretary of State James Buchanan instinctively shifted in his chair away from the president. Nicholas Trist, the State Department clerk that Polk had sent to Mexico City to get a few signatures and return, had taken it upon himself to decide the fate of the nation, to make decisions about territory and war that not even the president was authorized to make alone.

Trist wasn't the only person Polk was angry with. Santa Anna, who had proved that he was no great military strategist, was now showing that he was a fool as a leader, too. Polk had sent secret emissaries to Santa Anna, telling him that if he would stop short of Mexico City, stop fighting, and agree to the treaty that Trist had carried to Mexico, the United States would support him as the president of Mexico in the years to come. To Polk it was clearly the best opportunity for Mexico, offering the promise of years, if not decades, of stability. But Santa Anna had continued to fight, throwing away his chance of U.S. support.

"All I wanted was to settle the question of Texas and California," Polk insisted to Buchanan, but he was being disingenuous. Polk wanted to bring Texas and California into the Union, yet in a way that would bring admiration to his administration. Trist and Santa Anna were ruining a simple end to what up to that point had been a simple war.

"Can't that fool Trist see that Santa Anna has changed everything? Haven't you sent him messages telling him that the treaty is no longer valid?"

Buchanan shifted in his seat again. "Mr. President, we haven't sent Mr. Trist any instructions since we ordered him to return to the United States. We do not want to acknowledge his role there." Buchanan's reasoned response only made Polk angrier at the difficulty of the situation.

The next day, Buchanan returned with Jefferson Davis, the war hawk from Mississippi. Davis had left the U.S. Congress to join the fight in Mexico, but he had returned to make the case in Washington that Mexico now belonged to the United States. Davis was an outspoken supporter of slavery, and it was well known that he wanted to acquire more territory so that slavery could be expanded. But he surprised Polk with his argument. "Mr. President, we should set aside the question of slavery in the new territory for the time being. That is a question best decided outside of the heated atmosphere of war," Davis said. "What we need to examine is the temerity of the Mexican army, and their refusal to accept our generous terms. Our army has gained territory at a grave cost, and it is our right to keep the lands that Providence has delivered to us."

Polk was inclined to agree with Davis. But Buchanan added another compelling argument. "The security of our nation would be at risk if we were to now leave Mexico," he said. "Theirs is the least stable government in the world. There have been five presidents in the last three years, and there is no end in sight to the instability at our southern border. This present war will only increase the turmoil there. How long would it be before someone could come to power in Mexico by fanning the fires of resentment against the United States? Then there will be another attack against Texas, and we will be fighting this same war again *within your lifetime*."

Polk stared at the White House lawn. Buchanan was right, of course. Returning the captured territory to the Mexican government would be a decision that could haunt some future president. But Polk worried, was conquering a sovereign state something that he wanted to be remembered for? He continued to gaze toward the lawn. He must not worry about what future generations thought of him, he decided. He would be a part of the red Tennessee earth by then in any case. He was responsible for protecting the American people above all else, and the dangerous political situation in Mexico would constantly threaten U.S. citizens along the southern border. Davis and Buchanan had well spoken the concerns that had been present in his mind for days.

"I will order General Scott to arrest Mr. Trist. I want him transported back across the Rio Grande and released. If he crosses that boundary back into Mexican territory, I will have him shot as a traitor. Once he is back in Texas, do not offer him transportation back to Washington. From there he can walk back to Washington for all I care." Polk walked back across the office. "Mr. Davis, would you be willing to go to Mexico again, this time to deliver our present desires to whatever Mexican authorities you can find?"

Davis traveled to Mexico and presented the new treaty, which declared the U.S. border at one marine league south of Veracruz, extending westward to the 450-mile-long Balsas River, which empties into the Pacific

Ocean. The nation of Mexico would be less than one-fourth the size of its claimed territory before the war. With American troops occupying Mexico City, Guadalajara, Tampico, Veracruz, and most of the land covered by the treaty, there was little the government of Mexico could do to stop the Americans from establishing the new boundary. Mexico, which had been one of the largest nations in the world, one that before 1846 had been nearly twice the size of the United States, was now just another small Central American country.

But Polk's desire to protect the security of the United States by acquiring most of Mexico was never achieved. Adding the new territory to the United States brought numerous new conflicts, both expected and unimagined. The plantation owners of the Southern states began expanding the practice of slavery into the newly acquired territory, causing immediate turmoil. Radical abolitionists in Massachusetts, other New England states, and even New York began talk of secession. But the slave trade was stopped in Mexico by the Mexican Americans themselves. The plantation owners soon discovered that the Mexican Americans were even more strenuously opposed to the white men's practice of keeping people of a different color in subhuman bondage than were the U.S. abolitionists.

Debates soon broke out in Congress and across the nation on whether laws that specified the rights of citizenship to white Americans also protected the new Hispanic Americans. By the mid-1850s, America soon dissolved into a three-tiered legalized class system of whites, other people of color, and black slaves. It was a caste system bound to collapse.

In 1859, an army of nearly one thousand Mexican Americans wearing black masks attacked four slave

plantations near Monterey, killing the plantation owners, burning the plantations, and freeing the slaves.
James Buchanan, who by then had been elected president, refused urgent demands from white slave owners
across the South that U.S. troops be sent to the territory
to protect the white plantations from these Hispanic
"night riders."

The new states of Monterey and Veracruz had been
admitted to the Union as slave states in 1856, but within
two years, the people there had voted to prohibit slavery.
Buchanan repeatedly invoked "states' rights" as justifica-
tion for not overturning the voters' decisions in those
states. "All that the States have ever contended is to be let
alone and permitted to manage their domestic institutions
in their own way. As sovereign States, they, and they
alone, are responsible before God and the world for
whether or not slavery exists among them," he said.

But as the assaults against plantations spread as far
north as Dallas, Texas, white plantation owners of the Cotton States feared that the violence against them could soon
reach Mississippi and Alabama. They declared that if the
government would not protect their interests, they would
secede and form their own nation, the Confederate States
of America. They made good on their threat of secession,
and the result was a difficult and complex civil war. As the
Confederates fought against Union soldiers in the East, in
the states of Monterey and Veracruz, soldiers on both sides
had to fight against organized armies of Mexican-American revolutionaries who were fighting for independence.

President Abraham Lincoln had passed the Emancipation Proclamation in 1863 as a measure to help win the
Civil War, but as the conflict dragged on, on January 1,
1864, he proposed another radical measure. Lincoln pro-

posed returning all land south of the Rio Grande to the Mexican Americans, granting independence to the people of the former Mexican territory. But there was one caveat: The independence would come only at the end of the war. Although it was unsaid, Lincoln, who had been bitterly opposed to the Mexican-American War in 1846, was offering to trade independence for Mexico in exchange for a Union victory against the Confederate states. Across the new states of Monterey and Veracruz men lined up to join the Union army, and more than 240,000 men from the former Mexican territory soon joined the Union lines. By the summer of 1864 the Civil War was over. A war-weary nation reluctantly agreed to Lincoln's offer of independence for former Mexican territory, and on May 5, 1865, Mexico was remade as a major nation on the continent of North America.

Polk accepted the treaty Trist had negotiated, and fought for its passage in Congress. In the midst of the war, Ralph Waldo Emerson had predicted, "The United States will conquer Mexico, but it will be as the man swallows the arsenic, which brings him down in turn. Mexico will poison us." But Polk's decision prevented such a ruinous outcome.

Nicholas Trist returned to Washington in 1848 and found that he was unwelcome and unpaid. He was nearly broke from his months negotiating the treaty with Mexico, and when he returned to the United States, he took a clerical position with a railroad to make ends meet. It wasn't until twenty-three years later, when all was forgiven, that Congress paid Trist for his months of service negotiating an end to the Mexican-American War.

Polk's political whiskers had been attuned to the vibrations sent out by Zachary Taylor, because Taylor did run successfully for the presidency in 1848 as a Whig. Soon after, President Taylor sent William Tecumseh Sherman, who would later gain fame as a Civil War general, to the Southwest to survey the recently acquired territory. Sherman returned to the White House and reported, "I've been out there and looked them over, all that country, and between you and me I feel that we'll have to go to war again." Sherman paused, and added, "Yes, we've got to have another war."

President Taylor was stunned "What for?" he asked.

"Why, to make them take that darned country back!" Sherman answered.

Chapter 11

Lieutenant Colonel Lee's Decision

❦

Lt. Col. Robert E. Lee, the hero of Mexico City and Harper's Ferry, was offered the command of the U.S. Army days before being offered the command of the rebel forces. What if he had accepted Lincoln's proposal?

HIS life was a "desert of dullness," Robert E. Lee wrote in a letter in 1860. "I must stop and look to my tent for there is a dust storm raging that sifts through everything and clogs my pen while I write," he wrote. "The thermometer is 99 on the north side of my tent in a stiff breeze."

Lee was stationed at the border between Texas and Mexico with orders to protect the settlers from western outlaws, adventurous Comanches and Mexican bandits. Lee was fifty-four years old, having served in the military for twenty-two years. Despite his high ranking at West Point and his valorous service in the Mexican-American War, he was only a lieutenant colonel. He had been trained as an engineer, but he had transferred to the cavalry after he was passed up for promotion. Life in the cavalry wasn't nearly as eventful as he had hoped, however. When he wasn't baking under the hot canvas or chasing local hooligans, Lee spent his days evaluating military livestock. Secretary of War Jefferson Davis had sent thirty-two camels to the Texas desert to be evaluated as a replacement for mules in the Southwest. It was, Lee thought, an undignified end to what had been a distinguished military career.

Lee was rescued from his boredom by orders from Gen. Winfield Scott, which instructed him to report to Scott in Washington at once. Those hot desert days would be the last moments of boredom that Lee would experience for many years.

⬥

Lee was the son of Henry "Light Horse Harry" Lee, who had a successful, but equally ignoble, career in politics. (Henry Lee is perhaps best known for his eulogy of George Washington: "First in war, first in peace, first in the hearts of his countrymen.") In 1794, Light Horse Harry Lee was the governor of Virginia, and when President Washington requested assistance from the Virginia militia to put down the Whiskey Rebellion in western Pennsylvania, Lee rode off in pursuit of the rebels himself. Commanding fifteen thousand troops, he captured the rank of general but not a single rebel. Meanwhile, the citizens of Virginia were angry that Lee had abandoned the business of their state to do the bidding of the federal government, and while Harry Lee was absent, the voters declared the governor's office vacant and elected another man to the job.

Later, Lee added to his poor reputation in Virginia by losing the family fortune on land speculation and gambling debts. The Lee family's embarrassments didn't end there. A year later, young Henry (Harry's son and Robert's older brother) completed the family's downfall by losing the Lee family plantation to debtors.

Robert E. Lee took just the opposite approach to life from that of his ne'er-do-well father and brother. With no money to pay for an education, Lee entered West Point, where he excelled. He finished second in his class, but more impressively, he became the first person ever to make it through the academy without receiving a single demerit for inappropriate conduct.

Lee became an engineer and married Mary Custis, the great-granddaughter of Martha Washington. He gained not only a wife but also a plantation, Arlington Heights, which sat on the banks of the Potomac across the river from the District of Columbia. Arlington Heights was a magnificent five-thousand-acre estate that also was home to 150 slaves.

In the Mexican-American War, Lee served under Gen. Winfield Scott, who took an instant liking to the Virginian. In the war Lee bravely went behind enemy lines to map the land and provide intelligence. During one of his reconnaissance missions, he was surprised by a Mexican patrol at a spring. Lee hid under a log as the Mexican soldiers spent the afternoon filling their canteens, watering their horses, and even sitting on the log that sheltered him. Lee's brave forays behind Mexican lines led directly to two U.S. victories and were useful in several others. General Scott described Lee's actions as "the greatest feat of physical and moral courage performed by any individual in my knowledge" and proclaimed Lee "America's very best soldier."

Lee returned to Virginia, and out of frustration over not being promoted to general, he requested a transfer to cavalry. It wasn't long before he saw action in his new role. In October of 1859, a militant abolitionist named John Brown and a group of armed men seized the town of Harper's Ferry, Virginia, and the federal arsenal and armory located there. Brown and his men hoped to lead an armed insurrection of slaves. People in the South remembered all too well Nat Turner's rebellion in 1831, when fifty-five white Southerners had been killed and twenty-one blacks had been executed for the uprising. The entire nation focused on the events unfolding at Harper's Ferry. Led by Lee, marines and the Second Cavalry quickly put an end to the uprising, and Lee became a national hero.

❧

As Lee traveled to Washington, he didn't know the specifics of what Scott wanted to discuss, but he knew that Scott would press him to remain in the Union army even if his home state of Virginia decided to secede. Before Lee could present himself to General Scott in Washington, however, events elsewhere changed the nature of the discussion. On April 4, 1861, P. G. T. Beauregard, who commanded a group of South Carolina rebels, decided to prevent the federal Fort Sumter from receiving food. President Lincoln informed the governor of South Carolina that he would resupply the island fort by ship, but before the ships could arrive, the South Carolinians began bombarding Fort Sumter with cannon. On April 14 Fort Sumter surrendered, and the next day Lincoln announced that he would send seventy-five thousand troops to the South to ensure "the laws to be duly executed."

The crisis that had been building for decades over slavery and states' rights had finally exploded. "As far as I can judge by the papers, we are between a state of anarchy and civil war. May God avert us from both," Lee wrote. "I see that four states have declared themselves out of the Union; four more will apparently follow their example. Then, if the border states are brought into the gulf of revolution, one-half of the country will be arrayed against the other. I must try and be patient and await the end, for I can do nothing to hasten or retard it."

Not everyone agreed with Lee that there was nothing that he could do to hasten or retard the conflict. The *Alexandria Gazette* wrote an editorial expressing the importance of Lee's decision to his home state.

> *It is probable that the secession of Virginia will cause an immediate resignation of many officers of the Army and Navy from this State. We do not know, and have no right to speak for or anticipate the course of Colonel Robert E. Lee. Whatever he may do, will be conscientious and honorable. But if he should resign his present position in the Army of the United States, we call the immediate attention of our State to him, as an able, brave, experienced officer—no man his superior in all that constitutes the soldier and the gentleman—no man more worthy to head our forces and lead our army. There is no man who would command more of the confidence of the people of Virginia, than this distinguished officer; and no one under whom the volunteers and militia would more gladly rally.*

Virginia was a critical state in the dispute. Although it was a part of the South, it did not have complete sympathy for the Cotton States of the Deep South. Virginia was the most populous state in the South, and it had been the home of patriots George Washington, the nation's first president; Thomas Jefferson, the author of the Declaration of Independence; and James Madison, the father of the Constitution. More important was the industrial power the state held. Virginia was responsible for a third of all the goods manufactured in the South. Twenty percent of the railroads in the South were in

Virginia. The state was the home of many of the U.S. military's most important facilities. The rifle works at Harper's Ferry were one of the two military weapon manufacturing sites, an ironworks in Richmond was the only site in the South capable of manufacturing heavy ordnance, and Virginia was home of the Gosport Naval Base, one of the U.S. Navy's premier ship-building and repair facilities.

When the legislature of Virginia debated the question of secession on April 17, emotions were high on both sides. One person in favor of staying in the Union broke down into hard sobs in the middle of his speech; another legislator in favor of seceding was reported to have "wept like a child." Many people in Virginia, like Lee, were not convinced that war was necessary. "I am not pleased with the course of the 'Cotton States,' as they term themselves," Lee had written to his wife before he left Texas. "In addition to their selfish, dictatorial bearing, the threats they throw out against the 'Border States,' as they call them, if they will not join with them, argue little for the benefit. While I wish to do what is right, I am unwilling to do what is wrong, either at the bidding of the South or the North. One of their plans seems to be the renewal of the slave trade. That I am opposed to on every ground. . . ."

The next day, as the issue of secession was still being discussed in the Virginia legislature, Lee met with General Scott in Washington. Scott told him that a large military force was about to be sent into the South to quash the rebellion, and he said that Lincoln himself had asked that Lee command the mission. There was some indication that Scott never intended Lee to actually fight against his native South—General Scott's plan was to assemble such a powerful army, a display of overwhelming force, that cooler heads would prevail in the South and war could be avoided. But Scott was offering the lieutenant colonel the opportunity to assume command of the Union army.

According to witnesses, Scott quickly got to the brass facts: "These are times when every officer in the United States service should fully determine what course he will pursue and frankly declare it," Scott said.

Lee said nothing, so Scott continued. "Some of the Southern officers are resigning, possibly with the intention of taking part with their States. They make a fatal mistake. The contest may be long and severe, but eventually the issue must be in favor of the Union."

Again, Lee did not respond, and Scott terminated the interview. "I suppose you will go with the rest. If your purpose is to resign, it is proper you should do so at once. Your present attitude is an equivocal one."

As Lee rode home from Washington, he could see the white columns of Arlington for most of the ride. That evening he learned that the Virginia legislature had voted to secede, which meant that the issue of secession would be put to a voter referendum in late May. That night, Lee walked the halls of Arlington trying to sort out his thoughts and feelings. After midnight, he sat down and wrote his reply to General Scott's offer: "I have the honor to tender the resignation of my commission as Colonel of the 1st Regt. of Cavalry. Very resp'y Your Obedient servant, R. E. Lee, Col 1st Cav'y."

WHAT IF?

What if Robert E. Lee had decided to lead the Union troops?

A young black slave woman brought in Lee's supper and set it on the table in front of him; Lee didn't notice her or the food. He sat gazing into the flame of a candle. There is little in life as unsettling as not having a thought of what tomorrow might bring, and from the time that Lee had ridden back from Scott's office, staring at the columns of Arlington all the way, his mind offered no solace on that very issue. Two hours later one of the black slave butlers came in to remove Lee's plate, which had gone untouched. The young man stepped back, staring at Lee.

Lee looked up at him but didn't say a word. The young man finally found his courage. "Col'nel Lee, you look like you saw a ghost!" Lee lowered his eyes and nodded, and then turned his focus back to the candle flame.

Lee had in fact seen a ghost, but not as the young slave imagined. After so many years of trying to be as different from his father as possible, trying to make sure the hurtful things said about his father that had caused him such embarrassment would never be said about him, here he was, face-to-face with his father's decision. There was a story about his father that Lee had heard more times than he cared to recall. Lee didn't think the story was true, but he had heard it repeated so many times that he was never sure. According to the tale, Light Horse Harry had visited a friend, who lent Lee a horse for the evening for his return home. The friend had sent along one of his slave servants on a second horse so that Lee wouldn't be troubled to return the first horse later. Weeks later the servant returned after obviously spending a long journey on foot. When the friend demanded to know where he had been, the slave said that Harry Lee had sold both horses. When the friend asked why the slave hadn't returned home immediately, the poor soul replied, "Because General Lee sold me, too."

No one had ridiculed Robert E. Lee as they had his

father, but despite all the years, all the miles that he had marched, all the moments of temptation when he was able to contain himself knowing that his father would not have, now his life came to this. He faced the very decision that his father had faced years before, and for the first time, on this night, Lee finally understood at least this one moment in his father's life.

He had sworn an oath to the Constitution, and, like his father, he was going to uphold that oath. He could not betray his country, even if it meant abandoning the state that he loved.

The next morning Lee rode directly to the War Department offices and delivered his decision to General Scott. "Colonel Lee, it's like you've come back from the dead! When you left here yesterday I was convinced that I would never see you again," Scott said, rising from behind to desk to clap Lee on the back. Scott was smiling broadly. "I will report this to the president, and we will arrange for you to lead the army into the rebel territory as quickly as you can be ready. I have no doubt that this will be as simple as the episode at Harper's Ferry. As soon as your men fix bayonets and approach, the rebels will disperse. You should be back at your home by harvest."

Lee was naturally buoyed by Scott's reaction to his decision to head the Union army, but he didn't share Scott's view of the situation. On the way home, he stopped after crossing the river and looked again at the white columns of Arlington. If the people of Virginia did vote to secede, Lee would soon be leaving Arlington Heights for perhaps the last time. The irony that, like his father, he would be losing his family's plantation was not lost on him. Lee knew that there was a difference in losing the estate to debts and losing it over what he considered an honorable decision, but the result would be the same.

Although Lee had told no one other than General Scott and his wife about his decision, he was not surprised the next day to discover that all of the newspapers in the state had the news in headlines across the front pages. The papers also carried the news about the debates over secession in the Border States. "I must try and be patient," Lee told his wife, Mary. "For I can do nothing to hasten or retard the states' decision one way or the other."

Lee was being both modest and naïve. If the most able soldier in Virginia would not side with the Confederates, men were telling one another on street corners and in general stores, what kind of foolishness would we be getting ourselves into if we voted to join the rebellion? When the referendum vote was taken on May 20, the wealthy elite of Virginia, who expected the vote to be in favor of secession by an overwhelming margin, was given a shock. The voters rejected the legislature's decision by a vote of two to one. Virginia would stay in the Union.

Despite the furor over Virginia's decision to remain in the Union, three days later Arkansas joined the Confederate states of Texas, Louisiana, Mississippi, Alabama, Georgia, Florida, and South Carolina. But Arkansas was the only state to join the Confederacy after the voters of Virginia made their will known.

In North Carolina, most people were for rebellion, except for the Northern sympathizers living in the Appalachian regions. (In North Carolina, as in Virginia, Tennessee, and Georgia, the people living in the mountains didn't own slaves and were fiercely pro-Union.) But Virginia's decision to remain in the Union influenced North Carolina's decision in a major way. Instead of being just one of many states in the Confederacy, North Carolina would be at the front line of the war. Every time an army marched out of the north through Virginia, it would pass through North Carolina, and many people feared that by the end of the war there would not be a shop, farm, or village left standing. For that reason, on May 23 North Carolina also voted to stay in the Union.

In early June, Tennessee narrowly voted to stay with the Union, and soon after, the slave states of Delaware, Maryland, Kentucky, and Missouri also decided, despite their Southern sympathies, to side with Lincoln and Lee.

All of the other Border States followed Virginia's example to stay in the Union, except for Arkansas, where cotton was a major crop. Although each of the Border States contained thousands of Confederate sympathizers, instead of a civil war of fifteen slave states against fifteen free states, the fury of the nation would focus on the eight Cotton States that formed the Confederacy.

Lee quickly assembled fifty thousand troops into a striking force. He marched his men through the North Carolina foothills of the Appalachian Mountains, where his army was least likely to be disturbed by angry Southern sympathizers. In northern Georgia, Lee and his troops met Confederate general Joseph Johnston, Lee's former classmate at West Point, in the first battle of the war near Stone Mountain on June 17, 1861. After two days of vigorous fighting, Lee was able to push into Atlanta. With the quick Union victory, Lee now realized that he had the opportunity to turn the war from a protracted conflict into a quick war of suppression. Lee paused in Atlanta to use the city's railroad lines to add sixty thousand more men and boxcars filled with supplies. Three weeks later, the enormous Union army, the largest that had ever been assembled in North America, began marching directly toward the capital of the Confederacy, Montgomery, Alabama.

Lee met Confederate general P. G. T. Beauregard outside of Montgomery at the town of Opelika. Lee had an advantage in heavy armament, and he used it skillfully, pounding Beauregard's lines for three days. When the Union army finally made its charge on September, it ran through the Confederate lines like "a penknife through a newspaper," as Lee famously put it. Lee formally accepted the surrender of Jefferson Davis two days later, on July 12, 1861.

Had events stopped there, Lee's fortunes would have been quite different, possibly leading to the White House. But President Lincoln was being pushed by the Radical Republicans in Congress to retaliate against the rebelling states. Partly to punish the Cotton States, and partly for purely humanitarian reasons, on September 1, 1861, Lincoln submitted his Emancipation Legislation to Congress. "The monstrous injustice of slavery deprives our republican example of its just influence in the world—enables the enemies of free institutions, with plausibility, to taunt us as hypocrites," Lincoln said. With the rebelling states excluded from voting, the legislation passed, and slaves across the South—in the Confederate states and in the Border States that had remained loyal to the Union—were free.

Riots broke out in several Southern Border States; many slave owners in the Border States believed that their loyalty to the Union had been betrayed. The Alexandria Gazette declared that the entire state of Virginia had been swindled by Lincoln and Lee, and compared the two men to riverboat con artists. "It would be in the interest of

General Lee to find residence in some state that would welcome him," an editorial wrote. "General Lee has betrayed the loyal patriots of his home state of Virginia, and the memory of that betrayal will last for generations." Lee was not cowed by the vague threats, but fearing for the safety of his family, he sold Arlington Heights and took up residence at West Point in New York, where he was the commandant for the next eight years. Lee never lived in his home state of Virginia again.

On May 23, 1861, Virginians finally went to the polls to decide the question of secession. It was almost a foregone conclusion. The legislature had already voted to secede, Richmond had been declared the capital of the Confederate states, and the nation's presumably most able military officer, a native of their own state, had sided with the rebels. The citizens of Virginia voted for rebellion by a margin of four to one. The vote split the state, literally. Most people in the mountainous regions in the western part of the state didn't agree with the decision to secede, and the state divided in half, forming a new pro-Union state, West Virginia.

For Lee, personal honor and his responsibility to his home state were paramount. Given the humiliation to the Lee name brought by his father, especially the embarrassment of the abandonment and loss of the Virginia governor's office, Lee was willing to give up the commission in the U.S. Army that he had held for thirty-six years to join the rebellion. The day after he made his decision, he wrote to his sister and her family, "Now we are in a state of war which will yield to nothing," Lee wrote. "The whole south is in a state of revolution, into which Virginia, after a long struggle, has been drawn; and, though I recognize no necessity for this state of things . . . in my own person I had to meet the question of whether I should take part against my native state. . . . I have not been able to make up my mind to raise my hand against my relatives, my children, my home." Perhaps Lee had forgotten that his sister's son was a captain in the Union army—Lee would be raising his hand against his relatives, as well as his nation.

Lee's Surrender at Gettysburg

℞

The Confederate Army of Northern Virginia, defeated at Gettysburg, had to retreat through Pennsylvania and Maryland to reach safety on the opposite side of the Mason-Dixon line. What if Union general George Meade had counterattacked and cut off Lee's retreat?

IN late June 1863, in the middle of the four-year-long American Civil War, Robert E. Lee had boldly marched his Army of Northern Virginia, the main army of the Confederates in the eastern theater of the war, into Union territory.

The Confederates had crossed the Potomac River from Virginia into Maryland via a pontoon bridge, a bridge laid across boats floating in the Potomac. On July 4, 1863, Union general William Henry French sent a cavalry troop to destroy the bridge so that the Confederates would not be able to retreat. The cavalry cut loose the pontoon boats so that they would float away on the current, and then burned the platforms on the shore so that another pontoon bridge could not be quickly constructed. The troopers then destroyed another bridge over the river that the Confederates were in the midst of building, and even captured a Confederate ammunition train and threw the freight into the Potomac.

Lee's foray into Union territory was what President Abraham Lincoln had hoped for for months—Lincoln was convinced that if the Confederate army should ever be foolish enough to go across the Potomac River and drive deep into the Union states, it would "never return, if well-attended to." It was soon apparent, however, that the Confederate army wasn't well attended to, and the chance to end the war in 1863 slipped away in the night. Two more years of bloody fighting would follow.

In the summer of '63, Lee thought that after two years of fighting it was time to force the end of the war in a final, climactic battle. The Union cause looked dim. The Confederates had won at the Battle of Chancellorsville in the lowlands of Virginia, when the Union army had turned tail and scrambled back across the Rappahannock River into northern Virginia, and Union general Ulysses Grant had been befuddled and stalled by the fortifications around Vicksburg, Mississippi, out west. Lee decided that if he could cross Pennsylvania and capture Baltimore in eastern Maryland, the nation's capital of Washington, D.C., would be surrounded, and the Union government would have to agree to Confederate demands of a separate nation.

Lee assembled seventy-five thousand men and began marching through Virginia's Shenandoah Valley toward the Potomac River and Maryland. As reports of Lee's movements reached Washington, Lincoln told Union commander Joseph Hooker that he should attack Lee before the Confederates made it across the Potomac. Instead, Hooker proposed that while Lee was in Maryland marching north, it would be a perfect time to capture the Confederate capital of Richmond. As Lee moved north, Hooker wanted to move south. Lincoln concluded that Hooker had become scared of Lee, and on June 28 he replaced him with George Meade.

Lee's army marched north through Maryland and into Pennsylvania, where one of the Confederate divisions heard of a supply of shoes in the village of Gettysburg and headed there. A Union officer scouting in the area saw that the town was served by several roads and surrounded by defensible ridges, and reported to Meade on July 1 that this was a good place to stage the coming battle. Both armies began rushing toward the town to take up the best positions, the Confederate army hurrying from the north, the Union army moving up to meet them from the south. Later that same day the first fighting started, and soldiers from both armies ran toward the sounds of the guns at Gettysburg.

In the second day of the battle, the Confederates tried to assault the Union lines at the tops of the ridges, but the Confederates were repulsed in several bloody attempts. The two armies lost thirty-five thousand men in the two days. But the battle was not yet over. That night, Meade proved that he was a capable battlefield strategist, telling the general commanding the center of the Union line, "If Lee attacks tomorrow, it will be in your front."

An attack on the center of the Union line was exactly what Lee planned, and he ordered James Longstreet and George Pickett to lead the charge. The two armies spent the morning of July 3 exchanging cannon fire, then suddenly the Union guns were quiet. Thinking that their artillery had knocked out the Union cannon, at midafternoon on July 3 rebel soldiers began moving toward the center of the Union line in an incredible mile-wide assault. The Union cannon had not been destroyed, but Union general Henry Hunt had stopped their artillery barrage, hoping to lure the Confederates into a trap. The Union ruse worked—a half-hour after it began, Pickett's charge was over. Fewer than half of the Confederate soldiers who attempted to breach the Union line were able to return to camp.

The losses during the three days of fighting at Gettysburg were severe. The Union Army of the Potomac had lost twenty thousand men, about a fourth of its fighting force; the Confederate Army of Northern Virginia had lost twenty-eight thousand men, more than a third of its soldiers.

That evening, Lee met with Gen. A. P. Hill in Hill's tent. Under the light of a single candle, they spread a map out on their knees and tried to find a route out of the northern Union states, through the mountains. The wounded and prisoners would be put on the wagons and sent directly over local roads to the Potomac pontoon boats at the town of Falling Waters.

The next afternoon the wagon train of wounded was ready to set off for Virginia. The wagon train was immense; it held twelve thousand wounded men on a parade of misery so long—thirty miles—that it took more than a day to pass any given point. In the early afternoon of the fourth, the obstacles for the retreating Confederate army worsened when a huge thunderstorm began to pound down. "The rain fell in blinding sheets; the meadows were soon overflowed, and fences gave away before the raging streams," wrote John Imboden, the Confederate general charged with leading the wagon train. "During the storm, wagons, ambulances, and artillery carriages by hundreds—nay, by thousands—were assembling in the fields along the road from Gettysburg to Cashtown, in one confused and inextricable mess."

Finally, on July 5, the wagon train reached the Potomac, and the soldiers discovered to their horror the work of the Union cavalry, the remains of the destroyed pontoon bridge. Worse, the Potomac had risen ten feet because of the rain, making it impossible to cross on foot. "Our situation was frightful," Imboden wrote. "We had . . . all of the wounded that could be brought from Gettysburg. . . . My effective force was only about two thousand, one hundred men and twenty-odd field pieces. We did not know where the Confederate army was; the river could not be crossed; and small parties of [Union] cavalry were still hovering around. . . . As we could not retreat further, it was at once frankly made known to the troops that unless we could repel the threatened attack we should all become prisoners, and that the loss of his whole transportation would probably ruin General Lee."

The fighting army, what remained of it, was in poor shape as well. They were running out of ammunition—one Confederate artillery soldier who was captured reported that his unit had only two rounds left. Two days later, on the morning of July 7, Lee's army joined the waiting wounded men and began trying to establish defensive positions near the bank of the swollen Potomac River. In such a perilous position, the Confederate army awaited Meade's impending attack.

⸙

On the afternoon of July 4, the Union signal corps officers reported to Meade that Lee's retreat had begun, and that day Meade issued General Order No. 68, congratulating his men on having "driven from our soil every vestige of the presence of the invader."

In Washington, President Lincoln was dismayed at Meade's assumption that the Confederacy was a sanctioned nation that had invaded the nation to the north. Furthermore, the message implied that Meade appeared willing to let Lee return to Virginia.

Meade did seem to be too willing to let Lee escape. On July 4 Union general Birney reported that Lee's army was in full retreat, and requested permission to attack. The orders were approved, and Birney began positioning his men to ambush Lee's army, but then, unexpectedly, Meade sent written orders that they were not to attack. Even worse, Meade left the gaps through the Allegheny Mountains open for Lee's retreat. Meade had incorrectly assumed that Lee had posted men to hold the high ground above the gaps, and so Meade neglected to send troops there. In fact, Lee had not posted any men in the mountain passes. The Confederates quickly noticed that the Union army wasn't engaged in the chase: "The enemy had pursued us as a mule goes on the chase of a grizzly bear— as if catching up . . . was the last thing he wanted to do," one Confederate said.

Meade had some reason for caution. The Battle of Gettysburg left the Union army in poor condition to fight. "Our men have toiled and suffered as never before," wrote Col. Rufus Dawes of the Sixth Wisconsin. "Almost half have marched barefooted for a week. I have not slept in a dry blanket or had dry clothing since crossing the Potomac before the battle."

Still, the Union army held an advantage of eighty-six thousand men to the Confederate army's fifty thousand, and many of the rank-and-file were ready to meet the gray coats again. "You do not know with what confidence this jaded-out army goes forth to the harvest of death," wrote one Massachusetts soldier.

On July 7 Lincoln was still confident that the battle could be won. "Now if General Meade can complete his work, so gloriously prosecuted thus far, by the literal or substantial destruction of Lee's army, the rebellion will be over," he said.

Meade responded with atypical frankness. "My army is and has been making forced marches, short of rations and barefooted," he wrote to Henry Halleck, Lincoln's general in chief. "I take occasion to repeat that I will use my utmost effort to push forward this army." Finally, on July 12 Meade wired

Washington, "It is my intention to attack tomorrow, unless something intervenes to prevent it."

Halleck responded, "You are strong enough to attack and defeat the enemy before he can effect a crossing. Act upon your own judgment and make your generals execute your orders. Call no council of war. It is proverbial that councils of war never fight." Despite Halleck's instructions, Meade did convene a council of war with his top commanders. Five were opposed to attacking the Confederates, and four were in favor of launching an offensive. Meade decided to postpone the attack at least one more day. The next day, one of Meade's generals escorted Vice President Hannibal Hamlin and Senator (and future vice president) Henry Wilson to the top of the steeple of the Funkstown Lutheran Church to overlook the field of the upcoming battle. "We shall have a great battle tomorrow," Meade boasted.

General Lee didn't agree with Meade's schedule. The Confederate army was running out of food and ammunition. Lee had been able to establish a strong position over the week of waiting, and to some extent he welcomed another battle. But when his scouts told him that the Union soldiers were still building fortifications, Lee announced, "That is too long for me; I cannot wait for that." Then, as an afterthought, he added, "They have but little courage!"

On July 12, Lee's men found a warehouse and, by tearing it down, were able to build enough boats that in just twenty-four hours they were able to construct another pontoon bridge across the river. On July 13, ten days after the fateful day in Gettysburg, Lee was able to begin his escape into Virginia. The new bridge was crowded with wagons and horses, but over intervening days the river had fallen enough that it could be safely crossed. Lee had the tallest men in the army enter the river at the ford and, interlacing their rifles slings together to prevent foot soldiers from being swept away in the current, form two lines through which the Confederates crossed the river.

That night Meade's soldiers heard the Confederate army moving, but Meade didn't think it was possible that the Confederates could retreat. When he finally realized that they were, Meade ordered a pursuit at 8:30 A.M. Meade's soldiers were able to catch up the rear of the Confederates, and there was a small battle with 125 Confederate and 105 Union casualties, but the rest of Lee's army was able to escape to the safety of Virginia. "It was a sad morning . . . when we went over the slight knoll and through the weak rifle pits—which the enemy had hastily dug—and realized that they had gone—vanished—and our faces were again directed toward Virginia," one Union soldier wrote home.

On the afternoon of the 14th of July, Union general G. K. Warren telegraphed the War Department, "[Lee's] Maryland Campaign is ended. Have sent to me . . . all the maps you can spare of [Virginia's] Shenandoah Valley. . . ." Lee's Maryland campaign was ended, but the Civil War was, tragically, far from its conclusion.

WHAT IF?

What if Meade had pursued Lee after the Battle of Gettysburg?

In his tent, Meade awaited his generals to review the events of the third day of battle at Gettysburg. As with all great battles, the day had been spent in a surrealistic mixture of numbing horror and heart-pounding exhilaration.

Meade also felt relief, relief that he had not failed in the first test of his battlefield tactics. This had been just his fifth day of command, and he had apparently won the world's most significant battle since Wellington's victory at Waterloo. Lee's attack had come exactly as Meade had said it would, in the center of the line, and the line had held. It was as simple as that. The line had held. And now it appeared that the battle was his.

It had come at a great cost. Thousands of his men lay motionless on the farmland surrounding this little town; thousands more still lay on the ground groaning and pleading for mercies. His men who weren't casualties were exhausted and hungry, and by no means ready for another day of fighting. But fight they must, Meade thought to himself, fight or die, because such is the terrible fate of a soldier.

The most difficult decision of the three days was now before him. Lee might attack the Union lines again—if he did, Meade was certain that it wouldn't be an attack on the center again. He might attack from the rear—his scouts had told him that the Confederate general Jeb Stuart had taken his division around to the rear of the Union lines, but there had been no word on what had happened. It was possible that Stuart was still out there, waiting for his chance to attack. He might be gathering strength, waiting for daylight for a flanking attack on the line's shoulder. If Stuart were going to attack, he would have done it to support Pickett's and Longstreet's charge. He would not have waited.

Meade had a sense that Lee was not going to attack for a fourth day and risk the total destruction of his army. Meade knew that if his own army was in such poor condition, Lee's Army of Northern Virginia had to be near collapse. He had seen the mile-wide line of gray coats fall like stalks of wheat before the thresher. There had to be tens of thousands of casualties. Lee had to be running low on fighting men and supplies.

As Meade sat thinking, still waiting for his generals, he realized that the only thing for Lee to do was to move. But where?

If he had any fight left in him, Lee might move his army toward Baltimore, or even Washington. Those cities were defended but, Meade knew, not well enough. That would have been the next step if Lee had won this battle. Would he dare march there anyway?

No. No, Lee was not foolish, Meade thought. Lee would most likely retreat back through Maryland and cross back across the Potomac into Virginia, if not the next day, then the day after.

Well, there were two things he must do, Meade decided. He had to ensure that Lee could not attack Baltimore, and he had to close off the mountain passes in Maryland to make sure that Lee's army never returned to Virginia.

As his generals filed in and took their seats around the table, Meade presented his plan. "Lee surely has posted men in the passes through the mountains in Maryland," Meade said, pointing to the positions on one of several maps laid out on the table. "Birney, move your troops before first light, around here, and march as quickly as possible to these passes and try to capture them.

"Hancock, on my signal tomorrow, take the new troops of Fifth and Sixth Corps and go through the wheat fields, here, and toward Devil's Den. If Lee is planning to mount a retreat, he will establish a rearward line here. That is what I want you to hit."

The next afternoon, on July 4, the Union signal corps reported to Meade that Lee was assembling his men and that a retreat had begun. Meade immediately gave the order to Hancock to launch the offensive. Hancock ordered the fresh troops of the Fifth and Sixth Corps to march back over the wheat fields. The troops surprised the survivors of Longstreet's charge from the day before. The Confederates quickly tried to form a line in the rear of Lee's army, but the Fifth and Sixth Corps soon had Longstreet's men running through the fields and woods for their lives.

Meade quickly ordered Union generals Buford and Kilpatrick to lead their fatigued troops back into battle. The Confederates were organizing a wagon train of ambulances, wagons, and artillery pieces for the march back across Maryland. Within minutes after the Union line began advancing toward the Confederate wagons, a monstrous thunderstorm broke open. The thunderstorm slowed the encroaching Union soldiers, but it brought the Confederate wagons and horses to a standstill in the mud and mayhem. As dusk approached early because of the dark clouds, the Union troops were able to penetrate the Confederate ranks. The confusion of frightened horses, screaming wounded, and rifle shots, combined with the flashes of lightning and crashing thunder, made for a horrific scene, one that Confederate general Imboden said was like "we were in the infernal regions."

Lee's fighting force was able to make its way out of Gettysburg and the neighboring Cashtown, but on July 6, Lee's troops were stopped in western Maryland. Birney's soldiers had found the mountain gaps undefended, and they quickly took up positions that allowed them to fire down on the trails, halting Lee's progress. That afternoon,

Meade sent three officers under flag of truce to Lee. Meade's letter explained to Lee that his artillery and wagons of supplies had been captured, that his wounded were now prisoners of war (being cared for by Union physicians and nurses), and, most important, that the pontoon bridge over the Potomac had been destroyed. There was no option but an honorable surrender, Meade's letter explained in respectful wording. Lee agreed. On the evening of July 6, Lee surrendered his command and his troops.

On July 7 word reached Lincoln in Washington and Confederate president Jefferson Davis in Richmond, Virginia, that on July 4, Ulysses Grant had achieved victory over the Confederate forces in the western theater of the war in Vicksburg, Mississippi. Within two hours, a second message arrived telling Lincoln that Lee had surrendered his army to Meade. Although the Confederate government in Richmond would not fall for three weeks, it was obvious on that afternoon that the two-year rebellion was over.

Why did Meade not finish up the business of defeating Lee's army after Gettysburg? One reason was that the armies of both sides still tried to execute warfare by the rules of medieval times—this was a war, after all, in which officers on both sides often wore swords into battle—and Meade held strongly to the medieval notion that the loser is the army that leaves the battlefield first. Meade had waited, in part, to make sure that the Confederates were leaving the battlefield and not just pulling back to regroup for another attack. Then he could claim the victory.

The Civil War would grind on for another two years with enormous loss of life on both sides. Hearing the news of Lee's escape, Lincoln went for a walk in the White House lawn accompanied by the secretary of the navy, Gideon Welles. "What does it mean, Mr. Welles?" Lincoln wailed. "Great God! What does it mean?"

After the war, Maj. Gen. Abner Doubleday put into words what it meant. An attack on July 4 "would have saved two years of war with its immense loss of life and countless evils. . . . It is perfectly absurd to suppose that the enemy would choose a position on the bank of a deep river for the purpose of fighting us."

Lincoln was so distressed by Meade's actions that he wrote the general a letter:

> *I have been oppressed nearly ever since the battles at Gettysburg, by what appeared to be evidences that yourself, and General Couch, and Gen. Smith, were not seeking a collision with the enemy, but were trying to get him across the river without another battle . . . you stood and let the flood run down, bridges be built, and the enemy move away at his leisure, without attacking him . . . I do not believe you appreciate the magnitude of the misfortune involved in Lee's escape. He was within your easy grasp, and to have closed upon him would in connection with our late successes, have ended the war.*

It is one of the most famous letters never sent. Lincoln decided that because he had not gone through the horrors of Gettysburg, he shouldn't judge the actions of those who had. He was not completely forgiving, though. When he later did meet up with Meade, Lincoln told his general that the episode had reminded him of "an old woman trying to shoo her geese across the creek."

Whatever the reason, the Confederates realized what had transpired. After the war, Confederate general Imboden wrote, "If [Union] Generals Buford and Kilpatrick had captured the ten thousand animals and all the transportation of Lee's army at Williamsport, it would have been an irreparable loss, and would probably have led to the fall of Richmond in the autumn of 1863. On such small circumstances do the affairs of nations sometimes turn."

Chapter 13

The Murder of the Vice President

❧

Vice President Andrew Johnson and other top government officials were supposed to have been killed on the night Abraham Lincoln was assassinated. What if the conspiracy had been successful?

GEORGE Atzerodt had a problem. His friends expected him to commit a crime, a horrible crime, a murder. He was thinking about his problem while sitting in a bar, trying to decide what to do. It was true that he wasn't the best example of an upright citizen—he was just an immigrant laborer who had failed as an apprentice to a blacksmith, a sometime boatman who every now and then picked up extra money by smuggling supplies across the Potomac to the Confederates, and too often a drunk. But he wasn't a murderer, although he had promised his new friends that he would become one.

A week before, he had been boastful. After drinking a few whiskey cocktails with a friend in his room, he had showed off his Bowie knife and dropped an intentionally dark hint that "if this fails, the other will not," referring to his revolver. But now, sitting in the bar of the Kirkwood, one of Washington,

D.C.'s fanciest hotels, running up a tab that he didn't intend to pay, he tried to collect the courage to carry out the crime.

It wasn't just that Atzerodt was expected to commit a murder: His friends expected him to kill the vice president of the United States.

⁂

Washington, D.C., during the mid-1860s, naturally focused most of its efforts on the war, but there was time for entertainment not only for the politicians and government officials but also for the many military men traveling through the city. Many people spent time at the theater, where they were able to see a handsome young actor, John Wilkes Booth. Booth was a celebrity in the city, and a wealthy young man, too; his thespian efforts paid him more than twenty thousand dollars a year.

Among those who frequented Ford's Theatre, where Booth often appeared, was President Abraham Lincoln. One night in the winter of 1865, Lincoln sat in his box seats overlooking the stage. Booth was playing the villain in a play, and whenever his lines called for him to say something threatening or harsh, he walked to the side of the stage nearest the president's box and delivered his lines directly to Lincoln, even going so far as to shake his finger at the president. "He looks as if he meant that for you," Lincoln's guest said to him, and the president had to agree, "Well, he does look pretty sharp at me, doesn't he?"

What Lincoln didn't know was that Booth had more in mind than theatrics. Booth had used his celebrity to recruit a motley band of hangers-on to help him with an audacious plan to kidnap President Lincoln.

Booth and his grubby gang often met at the boarding house owned by Mary Surratt. During the early part of the war, Surratt had operated a tavern that had served as a station for Confederate spies, and when she sold the tavern and began operating a boarding house, her skullduggery continued. Surratt, her son John, and her daughter Anna were all swept up by Booth's charisma, and mother and daughter both held some amount of romantic interest in the handsome actor. John Surratt tried to keep his sister away from Booth so that she wouldn't become implicated in the conspiracy. But there was little doubt that the twenty-one-year-old John was in with Booth to the end. "It seemed as if I could not do too much or run too great a risk," he said.

The Surratts weren't the only ones being led by Booth. There was also Lewis Payne, a veteran of the Confederate army who had fought—and been wounded and captured—in Pickett's charge at Gettysburg; David Herold, an unemployed laborer; Atzerodt; and a few other characters of dubious reputations.

In March 1865, Booth took this group to dinner at Gautier's, one of Washington's toniest restaurants, an almost unimaginable place for such a band of misfits. Over an after-theater midnight dinner of oysters in a private room, Booth laid out his plan to kidnap Abraham Lincoln. Booth wanted to have one of the men grab the president and throw him to the stage below, where Booth and the others would push him out of the back door, tie him up, and drive him to Richmond, Virginia, the capital of the Confederacy. Then they would demand an end to the war and recognition of the Confederacy, as well as the safe return of all Confederate prisoners of war.

Booth's friends may have been starstruck by Booth's celebrity and impressed by the surroundings, but they still thought his scheme was far-fetched. They proposed simply hijacking the president's carriage instead and redirecting it to Richmond. The plan lacked the melodrama that Booth clearly craved, but it would work.

On March 17, Booth saw an opportunity. Lincoln was going to attend an afternoon performance of a play called *Still Waters*

Run Deep, and Booth and his band planned to capture the carriage on the return trip and hijack the president and his entourage to Richmond. When that plan fell apart, Booth and his band came up with another, simpler scheme to simply grab the president and hustle him off to Richmond. One night the group hid along a roadside, and when Lincoln's carriage rode up, they stopped, surrounded it, and threw open the door. To the surprise and chagrin of the kidnappers, Lincoln was not inside.

Booth was becoming frustrated at his gang's ineptitude in capturing the president, and he began envisioning a darker mission. He first gave an indication of it one afternoon while walking through the White House grounds with Lewis Payne. At the time, anyone could walk up to the White House and knock on the door, and President Lincoln was known for agreeing to see citizens who wanted to present one problem or another. Booth mentioned casually that someone could go to the door, present a calling card, and if shown to the president's office, shoot him on the spot. When Payne didn't rise to the bait, Booth asked the veteran if he had lost his courage.

Payne wanted to please Booth, and they quickly put together another plan to kill Lincoln. The president had the habit of walking each afternoon to the War Department and returning to the White House at about the same time. One afternoon in late March, Payne hid behind a hedgerow with a pistol, waiting for Lincoln. The sidewalks were covered in ice, and when a heavyset man walked toward the president and a companion, Lincoln began joking, "Major, spread out! Spread out," as if they were walking on ice over a pond that might give way under the weight of the three men. Payne, crouched behind the bushes, heard Lincoln shouting "Spread out!" and thought that the president had somehow discovered him. Instead of standing and firing, Payne sat quietly to see what would happen. In just a moment, Lincoln and the major had walked on past, and the chance for ambush was gone.

Larger events elsewhere soon changed Booth's plan to end the war in the South's favor. On Palm Sunday, April 9, Robert E. Lee surrendered the Army of Northern Virginia to Ulysses S. Grant at Appomattox, Virginia. The war was over, but Booth's ambitious desire to achieve immortality by influencing the conflict remained. On Tuesday, April 11, Booth and Payne were part of a large crowd that gathered on the grounds of the White House to hear President Lincoln deliver a speech on his plan for the reconstruction of the South. "Let us all join in doing the acts necessary to restore the proper practical relations between these States and the Union," Lincoln said. He gave one example of what he had in mind: "It is . . . unsatisfactory to some that the elective franchise is not given to the colored man. I would myself prefer that it were now conferred on the very intelligent and those who serve our cause as soldiers."

"This means nigger citizenship," Booth said, his voice full of disgust, and he tried to bully Payne into shooting Lincoln on the spot. Payne, though not the brightest candle in the box, thought better of murdering the president in the midst of a crowd. Still, Booth declared truthfully as they left the White House grounds, "That is the last speech he will ever make."

That week Booth learned that both Lincoln and Grant would be attending a performance of a comedy at Ford's Theatre, *Our American Cousin*. This opportunity seems to have quickly hatched another scheme in Booth's mind. In the days following Lee's surrender, Grant was easily the most popular man in the Union, even more respected than Lincoln. Instead of killing just the president, Booth realized that he could kill Grant, too. But he could go even further—he and his men could kill *all* of the top leaders of the government. Perhaps Booth thought that with the rash of assassinations the government would be in disarray, the Civil War would be reignited, and South could prevail. But it is perhaps more likely that he was motivated by his own hatred for Lincoln and his desire for immortality for himself. When he learned later that Grant would be traveling instead of attending the play, he only modified his plan.

On Good Friday, April 14, Booth had his gang spread out throughout the capital. Someone, most likely John Surratt, was to kill Grant on his train after it left Washington. Booth also considered the murder of Vice President Andrew Johnson to be important, but for a different reason. Although he wasn't as popular in the North as Lincoln or Grant, Johnson had been the only Southern senator who didn't leave the Union to join the Confederacy, and he had served as the military governor of his home state of Tennessee during the conflict. To Booth, Andrew Johnson was a traitor. Booth gave the job of killing him to Atzerodt. Others were to kill Secretary of State William Seward and Secretary of War Edwin Stanton. Booth saved the leading role for himself—he would kill Lincoln. The curtain would rise on this drama shortly after 10 P.M.

Booth had concerns about Atzerodt, and that Friday afternoon he stopped by the Kirkwood Hotel to check on him. Atzerodt was out drinking, and so Booth left his calling card for Vice President Johnson, a mischievous act that would give rise to conspiracy theories involving the vice president later.

That evening, Booth went to Ford's Theatre. He quickly found that he was in luck: The man assigned to guard the outside door to the president's box was gone, most likely off having a drink. Booth jammed the door shut with a piece of wood and then waited for the cue that would tell him that the stage was clear for his exit. When he heard one of the actors pronounce the words, "Don't know the manners of good society, eh?" Booth stepped forward into the box, put his small derringer pistol to the back of the president's head, and fired.

At ten minutes after 10, two men approached the home of the Secretary of State William Seward on horseback. One of the men, Lewis Payne, dismounted and went to the door. Holding a package, he told the servant who came to the door that he was bringing medicine from Seward's doctor, to be delivered to Seward in person. The servant objected, saying that Seward could not be disturbed, but Payne marched past him and up the stairs. After more protestations that the secretary of state was asleep, Seward's son agreed to check in on him to be sure. Payne drew a pistol, pointed it at the son's chest, and pulled the trigger. Instead of firing, the gun, jamming, merely clicked. Payne smashed the weapon into the head of Seward's son, beating him until he was unconscious on the floor. Payne then ran into Seward's room and lunged at Seward with a knife, trying to slash his throat, but someone came up behind Payne and grabbed his arm at the last instant, causing the knife to cut Seward's cheek. Payne ran out of the room and was confronted by another man rushing up the stairs. Payne stabbed him in the chest and ran out to the horse that Davy Herold was holding for him. Together the two men rode off into the night.

Twenty minutes later, at the Kirkwood Hotel, there was furious pounding at the door of the room of Vice President Andrew Johnson. Johnson answered the door himself and found Leonard Farwell, the ex-governor of Wisconsin standing there. Farwell told Johnson that Lincoln had been shot just minutes earlier. Sometime that evening, Atzerodt had decided not to kill Johnson and had left the hotel, throwing his knife in the gutter.

Booth's conspiracy against the Union government had been only partially carried out—and with mixed results. Seward's wounds were not fatal. The men charged with killing Grant and Stanton had disappeared without committing any violence. Lincoln, of course, was dead.

Atzerodt had left the vice president's hotel and gone

home—not that it did him much good in the end. He was later hanged for participating in the conspiracy, as were Lewis Payne, David Herold, and Mary Surratt. Booth was tracked down to a Virginia tobacco barn, where he was shot either by his own hand or by an overzealous soldier. He was dragged from the barn and died shortly thereafter.

WHAT IF?

What if George Atzerodt had played his role in John Wilkes Booth's diabolical drama?

Immediately after the shooting at the Ford Theatre, Lincoln was taken, unconscious, to the nearby home of a Mr. Peterson. Soon after, the streets surrounding the house were filled with concerned faces, many of them black. Every one of the people outside the home was full of adrenaline, ready to burst off in some direction at a moment's notice if given orders. But there was nothing to be done, and the crowd stood nearly silent, waiting for some good word.

Leonard Farwell, the ex-governor of Wisconsin, pushed his way through the crowd to go back to the Kirkwood Hotel, where he was staying, to tell the vice president of the horrible attempt on President Lincoln's life. Farwell was accompanied to the Kirkwood by Maj. James O'Bierne, and when the two men turned the corner to the hotel, they could see that another crowd had already formed outside it, presumably to share the news of the shooting of the president. Farwell made his way to the doors, where a hotel employee grabbed his arm and, not recognizing the boarder, shoved him back down the steps.

"I have to see the vice president. I have urgent news," Farwell said. The hotel manager looked at him, apparently trying to gauge his intentions.

"He's dead," the manager said. "That's all I can say."

Major O'Bierne pushed his way forward. "What did you say?"

With the major suddenly appearing, the manager realized that these two men weren't curious onlookers. "Inside," he said.

In the hallway on the second floor, Johnson's body lay facedown, his feet still inside the room. The Oriental rug in the hallway was stained black under Johnson's head. Neither Farwell nor O'Bierne needed any additional evidence that Vice President Johnson was, in fact, quite dead.

The men hurried back to the Peterson house to relay the news. Several members of Lincoln's cabinet had gathered there, as had Speaker of the House Schuyler Colfax. Attorney General James Speed was in the house, and just after midnight on Saturday, the cabinet officers and Colfax quietly met with Speed to determine how to proceed. All of the men realized that the president pro tempore of the Senate should be the new president, but there was a momentary disagreement over who that was. Speed thought it was Daniel Clark, a senator from New Hampshire, but Colfax reminded Speed that Clark had been the president pro tempore in the previous Congress. Congress had met in a special session in March of 1865, and the Senate had elected a new president pro tempore, LaFayette Foster of Connecticut.

At one in the morning, Major O'Bierne and Speed awoke Foster from a sound sleep and told him that he must come at once to assume the duties of the presidency. A light rain had begun to fall by the time the men arrived back at the Peterson house. As Foster entered the room, he was nauseated by Lincoln's appearance. The area around Lincoln's right eye was swollen, and the right side of his face was a purple-red. His breathing was heavy and labored, and a physician simply shook his head at Foster's silent expression asking whether Lincoln could survive. At 7:30 in the morning, word spread through the crowd that Lincoln had died. Within minutes, Supreme Court Chief Justice Salmon Chase swore in LaFayette Foster as the president of the United States. "May God support and guide you," Chase said, putting his hand on the shoulder of a visibly shaken and confused Foster.

That Saturday morning, rumors blew through the streets of Washington like wind before a cataclysmic fire. By midday many people learned of the failed attacks on Seward. Some people said that the perpetrators were members of a division of Confederate soldiers who were still hiding near Washington, preparing to attack the entire city. Some thought that the murders were evidence of a pending British invasion. Otis Stone, a director at Ford's Theatre, said he didn't think Booth had killed the president—Stone was certain that the murder had been carried out by Union troops hoping to revive the war and that Booth had leapt to the stage to chase after the true killers. By Sunday, however, most of the rumors dissipated as first Atzerodt, then Herold and Payne were arrested and freely admitted their roles in Booth's conspiracy.

In the days after Foster's inauguration and the ornate funeral parade of Abraham Lincoln, many of the more radical Republicans admitted, quietly, that the process of Reconstruction was going to be easier now that Lincoln was gone.

For Vice President Johnson's death there was even

less false mourning. Johnson had never been well-liked in Washington; because of his impoverished childhood he resented the North's wealthy men of industry, and he did not attend social events in the city. He was also a Southerner and a Democrat. Worst of all, Johnson had held the peculiar view that because secession had been an illegal act committed by individuals, the states themselves existed as before and retained full constitutional rights. If Lincoln had been the only one killed and Johnson had become president—Thaddeus Stevens told Colfax that he literally shuddered at the thought—the task of Reconstruction, with appropriate justice handed out to the traitors of the South, would have been much more difficult.

Benjamin Wade, a senator from Massachusetts, had suggested to Lincoln before his death that the federal government hang a few Confederate traitors and then let the rest go without punishment, but the events of the week changed his opinion. The murders of the president and vice president, and the attempted murders of the secretary of state and General Grant, brought out a lust for revenge that most people in the North had theretofore contained. Wade reflected the views of many in the nation when he said of the leaders of the Confederacy, "I would arrest them. I would try them. I would convict them and I would hang them. Leniency for the masses—halters for the leaders!"

The revenge for the sins of the South brought a dark cloud over the latter part of 1865. Following the Booth conspiracy trial, which was a farce of justice, sixteen people were hanged for the murders of Lincoln and Johnson. Not only were all of those directly involved in the conspiracy hanged, but also Booth's brother Julius, who had been performing in a play in Cincinnati at the time; Samuel Mudd, who had set Booth's broken leg during his flee from justice; and Anna Surratt, who was convicted solely because investigators found photos of John Wilkes Booth hidden behind a mirror in her room.

The harsh measures didn't stop there. Jefferson Davis and Alexander Stephens, president and vice president of the Confederacy, were quickly tried and hanged for treason, as were many other top Confederate government officials. John Breckinridge, the Confederacy's secretary of war who had been the vice president of the United States just five years before, had fled to Cuba, but he was tried in absentia and sentenced to death should he ever return to the United States. At an enormous gallows built on the grounds of the infamous Confederate prisoner-of-war camp in Andersonville, Georgia, Robert E. Lee and dozens of other officers of the Confederacy were hanged. (This prompted the resignation of Ulysses S. Grant, who had promised Lee a general amnesty.)

The bloodletting ended by late 1865, but the punishment of the rebellious states had just begun. "The states that rebelled and took the lives of so many must never again be the equal of the states that remained loyal to the country that we hold dear," Foster said in an inaugural address two days after he was sworn in as president.

The eleven former states that had made up the Confederacy soon saw their recognition as members of the United States revoked. Radical Republicans, led by Colfax in the House of Representatives and Benjamin Wade in the Senate, eliminated the state lines and gave large portions of land to the states that had remained loyal to the Union. The states that had rebelled would exist no more. Most of Virginia, one of the thirteen original colonies, was given over to Maryland, making that new and formerly small state one of the largest east of the Appalachian Mountains; other portions of Virginia were given to the new state of West Virginia. Portions of Arkansas were given to Missouri, portions of Texas were given to the territory of Colorado, and Tennessee was absorbed completely by Kentucky.

The portions of the rebel states that remained were divided along new lines, constructed both to divide the loyalties of the old states and, to whatever extent possible, to benefit the Republican Party. The new districts were given the same rights, and limitations, as territories that had not yet been admitted to the Union. The Wade-Davis bill, which had been ignored by Lincoln when he was president and failed because of his pocket veto, was finally passed. The bill directed that after half of a state's white male voters had sworn an oath of loyalty to the Union and vowed that they had not aided the Confederacy, the state would be allowed to hold a constitutional convention and reapply for admittance to the Union. As with any piece of legal paper, the trick was in the details. Because no former Confederate state could meet the test—it would be nearly impossible for half of a state's white males not to have aided the Confederacy by serving in the military or holding some government office—none of the newly created Southern territories were admitted as states for a generation.

Martial law was imposed in the territories and maintained by a U.S. military that was almost as large as the Union army had been during the Civil War. For more than thirty years, most of the South remained under what amounted to severe military dictatorships. The presence of the military was certainly necessary, because the North's harsh treatment spawned resistance groups in the South eager to renew the war for Southern independence, and sporadic violence against federal troops was common. It wasn't until 1914, nearly fifty years after the end of the

Civil War, that the last federal troops left the South, and the divided nation was made whole.

Less than a week before Lee's surrender, on April 3, 1865, Union forces had overrun the Confederate capital of Richmond. The next day, while the city lay in still-smoldering ruins, President Lincoln visited Richmond and arrived at the door of a familiar house. It was the home of George Pickett, the Confederate officer who had led the failed charge into Union lines at Gettysburg. Pickett's wife opened the door holding a baby, and stood there, stunned. "I am Abraham Lincoln," he said, but of course she knew who he was. Besides being the president of the United States, he had helped get her husband into West Point before the war. "The president!" she said. "No, Abraham Lincoln, George's old friend," Lincoln said. The attitude that Lincoln showed toward George Pickett was the same that he planned to show to the entire South, that of a friendship renewed.

At the close of the war, Lincoln requested the incredible sum of $400 million for the reconstruction of the South. He was often nearly alone in his desire to rebuild the Southern states. A story, possibly apocryphal but probably not far off the mark, had Lincoln conducting his cabinet meetings and putting one of the issues of Reconstruction to a vote. All seven cabinet officers voted against Lincoln's plan. Lincoln responded, "Seven no and one aye. The ayes have it."

Andrew Johnson, although fiercely loyal to the Union and the Northern forces, was himself a Southerner. Following Lincoln's death, Johnson wanted to carry out Lincoln's plan of reconciliation and rehabilitation for the Confederate states. Johnson was impeached on trumped-up charges for his trouble, although he was eventually acquitted. (If Johnson had been convicted, the radical Benjamin Wade would have become president—Wade had even announced who would be in his cabinet before the Senate unexpectedly acquitted Johnson.)

Most others weren't so generous toward the South. Congressman Thaddeus Stevens, one of the leaders of the so-called Radical Republicans, wanted much more retribution. (Stevens was probably partially motivated by personal bias, because when Lee had marched to Gettysburg, some Confederate soldiers had ignored Lee's orders against destroying public property and had burned Stevens's ironworks.) Stevens and the other Radical Republicans were able to press for harsh Reconstruction legislation and attempted to override Johnson's vetoes. "There is a wild delirium among the Radical Members of Congress which is no more to be commended and approved than the Secession mania of 1860," wrote Lincoln's friend Gideon Welles in his autobiography. "In fact, it exhibits less wisdom and judgment, or regard for the Constitution, whilst it has all the recklessness of the Secession faction."

Every schoolchild knows that Lincoln kept the South from dividing the nation during the Civil War, but it was his generous policies, carried forward by Johnson and to a greater extent by the memory of Lincoln, that continued to keep the nation from being torn apart during Reconstruction by the Radical Republicans in Congress. Had the immigrant laborer George Atzerodt played his part in John Wilkes Booth's diabolical drama, the consequences for the country could have been devastating. It is unlikely that Booth would have succeeded in throwing the country back into civil war—few had much interest in that, not even in the former Confederacy. But many in the North would have wanted revenge on the vanquished South. The consequences for the former Confederate states—and for the nation as a whole—would have been severe.

Chapter 14

President Tilden

Democrat Samuel Tilden received three hundred thousand more votes than Republican opponent Rutherford B. Hayes in the election of 1876, and lost. What if Samuel Tilden had become president in 1877?

IN the late 1880s, in the midst of Grover Cleveland's first term as president, a small group of the most influential Democrats and Republicans in Washington gathered for an intimate dinner. During the evening the veteran pols began recounting the events of the famously corrupt election of 1876 a decade before. Finally a jovial President Cleveland threw both hands in the air and said, "What would the people of this country think if the roof could be lifted from this house and they could hear these men?"

One of the raconteurs probably spoke for the others when he said, "If anyone repeats what I have said, I will denounce him as a liar." One of the other guests at the dinner, Henry Watterson, who did write about some of the events of the election of 1876, said, "The whole truth underlying the determinate incidents which led to the . . . seating of [President Rutherford B.] Hayes will never be known."

This much is: In November 1876, Samuel Tilden, the Democratic presidential candidate, won the popular vote by three hundred thousand votes. But in March 1877, Republican Rutherford B. Hayes was sworn in as president in a secret ceremony at the White House.

In 1868, when Ulysses Grant had first run for president, he wasn't completely sure that he even wanted to be in politics—at one point he considered running for mayor of his hometown of Galena, Illinois, because he wanted a sidewalk from his house to the train station. But after eight years as president, Grant thought that perhaps he enjoyed the job so much that he'd run for a third term. Few others wanted more of a Grant presidency, because, unfortunately for Grant, who had arguably been the greatest general in United States history, his presidency had been a disaster.

There had been corruption, most notably the Crédit Mobilier scandal, where a company that relied on government contracts gave stock to congressmen and high government officials so that they wouldn't ask too many questions about how the company was being run. Making matters worse, Grant had appointed so many friends and cronies to important offices that Massachusetts senator Charles Sumner said that the administration suffered from "a dropsical nepotism swollen to elephantiasis."

After such a performance, both Democrats and Republicans were calling for governmental reform, especially civil service reform. Each party began searching its ranks looking for presidential candidates who favored reform and didn't have personal scandals. At their convention in Cincinnati, the Republicans rejected Grant and picked as their candidate a hometown hero, Ohio governor Rutherford Hayes, who had first gained fame as a Civil War general. Pointing out Hayes's integrity, the Republicans took as their slogan, "Hurrah for Hayes and honest ways!"

The Democrats, meanwhile, picked Samuel J. Tilden, governor of New York. The Republicans may have said they wanted governmental reform, but the Democrats meant it. When he was district attorney of New York, Tilden had broken up the notorious Tammany Hall and sent Boss Tweed and his henchmen to jail. Tilden had continued his reforming ways as governor, and in a campaign that was to be about governmental reform, the Democrats realized that not picking Tilden, who was not particularly well-liked, "would be like the play of *Hamlet* with Hamlet left out."

Although the Democrats lacked the Republicans' felicity with words—"Tilden and reform!" was their campaign slogan— most voters knew that when it came to real reform of the too-cozy relationships between government and the corporations, it was the Democrats and not the Republicans who would do the job.

Election day, November 7, 1876, was a dark day in American democracy. Across the South, blacks were harassed at the polls by Democrats and prevented by violence from voting. In Louisiana it was reported that "assassination [of black voters] for political opinions is so common that little seems to be thought of it." At some polls, illiterate blacks were tricked by ballots that were printed with Republican symbols that actually contained the names of Democratic candidates. The Republicans had their share of electoral assistance, too. In the areas in the South where federal troops (who were pro-Republican) were in place to ensure that blacks were allowed to vote, the soldiers did such a good job that many blacks were able to cast their votes for Republican candidates several times.

On that Tuesday evening, it appeared that Tilden had won the election. Mr. and Mrs. Hayes and a gathering of friends looked at the telegraphed election results with dismay. When the Republican didn't do as well as expected in his home state of Ohio, they assumed that this indicated a defeat for their party. Lucy Hayes excused herself from the gathering and went to bed with a headache; when Hayes joined her past midnight they consoled each other by pointing out that their lives would be much simpler outside of the White House. The next morning the Republican newspaper *The Indianapolis Journal* reported, "With the result before us at this writing we see no escape from the conclusion that Tilden and Hendricks are elected."

A few men in New York didn't see it that way. Coming home from a late party, the unscrupulous Republican Daniel Sickles (the former Union general) stopped at his party's headquarters and realized that although Tilden was winning the election, Hayes could still win if he carried three states in the South plus Oregon. Zach Chandler, the chairman of the Republican National Party, had fallen into a whiskey-induced slumber, so Sickles quickly sent a telegram to the Republican leaders in those states saying, "Can you hold your state? Answer at once," and signed Chandler's name. At 6 A.M. the editor of the *New York Times*, John Reid, rushed to the Republican headquarters to tell them that the Associated Press was reporting that Tilden had won the election. Waking Chandler, they sent more telegrams to the Southern states telling them that Hayes's election depended on them and warning the Republicans in these states that the Democrats might be doing something unethical (!). Later that day Chandler issued a statement saying that Hayes was the winner, an inaccurate interpretation Reid seconded in the later editions of the *Times*.

Tilden had won the popular vote (although with significant

help from the illegal activities) but had failed by one electoral vote to carry the election; Hayes fell twenty electoral votes short of victory. To assume the presidency, Tilden would need to win one of the twenty electoral votes in the states where the election was still in dispute because of irregularities at the polls. Hayes needed to win all twenty.

The battle was on. To ensure a favorable outcome, both parties sent armies of party officials to Florida, Louisiana, and South Carolina. Immediately trouble began. In Florida, the train carrying the Republican leaders was derailed outside of Tallahassee. Not long after, Tilden's nephew, working out of Tilden's house, tried to bribe election board members in the three disputed states. The electors weren't insulted—they indicated that they were willing to be bought. Meanwhile, in Louisiana, the chair of the election board, which had already pledged its electors to the Republicans, tried to blackmail his own party by insisting on a $250,000 payment to keep the pledge. When the Republicans refused, he offered the Louisiana election to the Democrats for $1 million. In the end, the electors did vote Republican after a payment of $200,000.

When Congress met on December 6 to count the electoral votes, a situation not anticipated by the Constitution presented itself. Louisiana and South Carolina each sent two sets of "official" election results, while Florida sent three separate sets.

As the politicians in Washington debated how to proceed, others in the country began their own machinations. The Democrats wanted Tilden to speak out about the election abuses, but one Democrat, after pleading with Tilden, walked out of his house and said in disgust, "Oh, Tilden won't do anything, he's as cold as a damn clam."

Others were willing to try a different tack, and most of the plans involved violence. "Tilden or blood!" became a rallying cry. In late November Grant had Washington reinforced with light artillery, and a Republican representative predicted that by March 4, the date of the inauguration, congressmen would be cutting each other's throats. Hayes's son Webb, who would later win a Congressional Medal of Honor during the Spanish-American War, began acting as an armed bodyguard for his father. In fifteen states the Democrats organized armed clubs of Tilden-Hendricks Minute Men, and there were rumors that the governors of some Southern states had selected a commander in chief to lead a rebellion. There was widespread concern that the regular army, which had dwindled to just twenty-five thousand men after the Civil War, would not be able to control any type of organized uprising. When the Democratic chairman asked William Sherman, commanding general of the army, what he would do if there were a conflict after March 4 and no new president was in place, Sherman said, "The term of President Grant ends at twelve o'clock on the fourth of March. He will then be in no position to give orders to me, and I shall receive no orders from him."

The political crisis in Washington was becoming a crisis for the entire country. Abram Hewitt, national chairman of the Democratic Party, wrote that "business was arrested, the wheels of industry ceased to move, and it seemed as if the terrors of civil war were again to be renewed."

Fortunately, cooler heads worked out another way. Some Southerners who were eager to return to home rule (i.e., white rule) realized that Tilden, the governor of New York, might not be any more sympathetic to their cause than Hayes. They, of course, held the key votes for the presidency, and they began meeting in secret with representatives of Hayes. Southerners weren't the only ones worried about the election outcome—Northern businessmen were worried that if Southerners were angry over a Tilden election, the discord might hurt their dealings in the region. Finally, in a late February meeting in

Washington's Wormley Hotel, a group of Ohio Republicans acting on behalf of Hayes met with a group of Southern Democrats from Louisiana and South Carolina. A deal was struck that would give Hayes the election. In exchange, the Hayes administration would remove federal troops from those two states, appoint a Southerner to the cabinet, and allow the Southern states to create their own laws without interference from Washington.

On March 1, 1877, Hayes left for Washington in order to be present in the capital should he be the one chosen as president. When he left his home of Columbus, Ohio, he told the crowd that had assembled to see him off that there was a chance that he would "be back immediately." It wasn't until the next day, as his train was traveling through Pennsylvania, that Hayes received final word that he had received enough electoral votes to be inaugurated as president. Because of concerns about his safety, Hayes was inaugurated two days later in a secret ceremony in the White House. Two days after that, a public inauguration ceremony was held without disruption.

WHAT IF?

What if Hayes had refused to accept the deal offered by the Louisiana and South Carolina Democrats to deliver their electoral votes to him in the election of 1876?

"Rutherford B. Hayes *be* damned!" a portly man rushing out of the front door of the Wormley Hotel said as he brushed past a young man trying to make his way into the hotel.

The young man, "Johnny" Apple, was a just a first-year cub reporter for the *New York Times*, and he was working on the story of a career. A friendly bartender at the Wormley had tipped him off that a meeting of the strangest sort was taking place in a suite of rooms upstairs in the hotel. A group of prominent Republicans from Ohio, including James Garfield, was meeting in the suite occupied by Stanley Matthews. The bartender, Chet, had whispered to Apple that a second group of men had entered the hotel shortly after the Ohioans, but these men were wealthy white Southerners. Chet had noticed that when the Ohioans were milling about in the lobby, so were the Southerners.

"I've seen some surprising combinations in this town," Chet said as he exercised a dish towel against the side of a glass. "And more than a few men have told me secrets about goings-ons in this city. I used to say that nothing could surprise me anymore."

Chet, who fancied himself a natural storyteller, set down the glass and leaned toward Apple for dramatic effect. "But if I live to be as old as Methuselah, I never would have thought that I would see sober, white Southerners meeting in a room with a bunch of Republicans."

Apple feigned nonchalance, running his finger around the edge of a glass of bourbon that he had been nursing for nearly ninety minutes. "Any of 'em happen to say any names?"

"Umph," the bartender grunted, turning around to stash the clean glasses behind the bar. Something was most definitely going on, Apple knew, because all afternoon, any conversation ended as soon as he asked who was upstairs. Apple spent the next three hours sitting in the bar, trying to buy as few drinks as possible and not get thrown out of the hotel. He knew that sooner or later a herd of politicians was bound to get thirsty and head for the nearest watering hole. Like a lion, Apple knew that all he had to do was to identify the weakest member of the herd, separate him from the others, and then devour him.

Apple had found his man, a Louisiana Democrat deep into his fourth bourbon. "Listen here, son," said the pol. "Those soft-headed, hymn-singing, lemonade-drinking, holier-than-thou Republicans are the worst thing that ever happened to that party." The wait had been worthwhile. "And then you have them New York reformers. They're just as bad as the Republicans. Why is it, every twenty years or so some New York do-gooder gets himself elected and begins putting crooks and greedy civil service workers in jail, and then struts around like the rooster that thought he made the sun rise each morning? Well, hell's bells, if you excuse my language, how hard is it to find a crook or a thief in New York? You could stand on the street corner and swing a switch around and hit three of 'em.

"We're in a fine mess now. With Mr. Hayes we got ourselves a Bible-reading Methodist and with Mr. Tilden we got ourselves a New York reformer, and there ain't none of them that will sit down and listen and make a deal. If we don't do something soon, we're heading for one big train wreck. I tell you, son, the caliber of politicians today is just shameful."

By midnight, Apple had learned much of what had happened in Matthews's suite upstairs in the Wormley. Hayes, working through fellow Ohioan James Garfield, a U.S. congressman and a member of the electoral commission, was refusing to accept the Southern Democrats' terms for fear that the Southerners would trample on the rights of blacks in the region.

The next day, Apple telegraphed this story to his editor at the Times:

WASHINGTON, D.C.—The fortunes of Republican presidential candidate Rutherford B. Hayes of Ohio disappeared in the night at Washington's Wormley House hotel earlier this week. Representatives of Mr. Hayes conducted meetings in the hotel with members of the Democratic Party from Louisiana and South Carolina in which the Democrats presented Hayes's Ohio comrades with a plan that would end Reconstruction in the rebel states in exchange for the necessary electoral votes to secure election. However, those Republicans representing Mr. Hayes said that he simply refused to consider such an affront to democracy. . . .

Apple's editor at the *Times*, John Reid, nearly fell out of his chair when he saw the story that one of his most inexperienced reporters had submitted to the paper. The story, if it had run, would have meant that any effort to secure a deal between the Republicans and the Southern Democrats was finished.

Reid didn't know that the deal was finished anyway. Southern Democrats had failed to sway the righteous Republicans, but Democrats in Washington were confident that they could convince Northern businessmen. Concerned about what a Hayes election and continued federal occupation of the Southern states would do to their sales to the region, the next evening several prominent businessmen invited three of the Republican members of the electoral commission, which was to decidethe election, to a second meeting in Washington's Willard Hotel.

Leading the meeting were several executives of the railroad companies. It wasn't unusual for these men to meet to discuss business, even though they all worked for different railroad lines and were, technically, competitors. On this day, however, they wanted to impress on some of their fellow Northern Republicans that it was time to end Reconstruction and that Tilden and Hendricks were the men to do it.

The railroad men reminded the Republican electors that the Gadsden Purchase of 1853, which had bought the southern portion of New Mexico from Mexico, had been made so that a transcontinental line could be built across the southern portion of the United States. This line would open up new markets in Texas, Arizona, New Mexico, and southern California for goods produced in factories in the Northeast. A Southern line would have several advantages over the current rail line, the most obvious being that it would not be closed because of winter snowstorms. Such a railroad line could not be built without the approval of several Southern states, but that approval stood little chance of happening as long as federal troops were stationed around the statehouses in the former Confederacy.

Everyone uses the tools they are familiar with. For attorneys this means litigation; for politicians this means legislation; but businessmen have one of the best tools of all. The railroad men proposed forming a new company to build the rail line, and the construction of the new line would be subsidized by the federal government. Unlike the Crédit Mobilier escapade, this new company would be on the up-and-up, at least in its use of the federal government's money. In exchange for the votes needed to give the election to Tilden, who would be expected to end Reconstruction, the spouses of each of the members of the commission would be given an opportunity to purchase stock in the new railroad company. If the spouses lacked the money, loans could be arranged using the stock as collateral.

The next day, the electoral commission elected to give the presidency to Tilden by a vote of ten to five. The three Republicans who switched their votes from Hayes to Tilden claimed that they did so for the noblest of reasons, citing the threat of violence, the damage the uncertainty was having on the nation, and a sense of fair play, as Tilden had clearly won the general election. Many people in the nation were suspicious of the sudden turn of events, especially because of the high-minded reasoning of the Republican electors who had switched their votes, but no details of the Willard Hotel arrangement would leak out to the press until the turn of the century.

With federal troops assembled about the platform to dissuade any critics of the electoral commission's decision, Samuel Tilden was sworn in as president on March 5, 1877.

As president, Tilden discovered that civil service reform was much more difficult than he had anticipated. But he found more acceptance for his plans to end Reconstruction. Tilden had planned to remove federal troops from the Southern states and to return the governance of the states to their citizens instead of Republican appointees. But some people in both the North and the South thought that this end to Reconstruction didn't go far enough, among them Tilden's vice president, Thomas Hendricks.

Hendricks was a racist; he considered blacks inferior, and while a U.S. senator he had even served as the attorney for a racist group, the Knights of the Golden Circle. Hendricks, who was known for his political shenanigans, began speaking forcefully in favor of Tilden's plan to end Reconstruction. In appearances before groups in both the North and the South, Hendricks began calling for the repeal of the Fourteenth Amendment to the Constitution. Certain articles in the Fourteenth Amendment did codify the Reconstruction policies, specifying that anyone who held any type of position in the Confederacy could not hold any office in the government, whether as an elected official, as a civil service employee, or in the military. Clearly Reconstruction could not end until the Fourteenth Amendment was repealed, Hendricks said, knowing too well that the amendment also gave full rights of citizenship to any person born in the United States. By 1879, both the Fourteenth Amendment and the Fifteenth Amendment—which said that "the right of citizens of the United States to vote shall not be denied or abridged"—were repealed. Unfortunately for the nation's black citizens, a revocation of civil rights turned out to be something that many people in the North and South agreed on.

The governor of Ohio, James A. Garfield, had warned Hayes not to accept the deal worked out in the Wormley Hotel. "A compromise like this is singularly attractive to that class of men who think that the truth is always halfway between God and the Devil, and that not to split the difference would be partisanship."

Hayes did accept the brokered presidency, however, and one reason may have been a genuine concern for civil rights in the South. Allowing home rule in the South meant that enforcement of civil rights for blacks was no longer the concern of the federal government. "The Negro will disappear from the field of national politics," the *Nation* correctly predicted. Hayes had agreed to end Reconstruction, which had bitter consequences for the nation. But by doing so he was able to preserve the basic pillars of civil rights in the United States.

Hayes's decision to allow home rule in the South ushered in the odious "Jim Crow" laws that legalized discrimination. But he accepted the deal in order to protect civil rights, not to diminish them. Hayes expressed concern several times that the Fifteenth Amendment to the Constitution might be overturned if Tilden and Hendricks were elected. Hayes promised black abolitionist and writer Frederick Douglass that "for the protection and welfare of the colored people the Thirteenth, Fourteenth, and Fifteenth amendments shall be sacredly observed and faithfully enforced."

At the end of Hayes's single term as president, even Democratic chairman Hewitt admitted that the Hayes presidency "was creditable to all concerned and was far better than four years of unrest which we should undoubtedly have had if Tilden had occupied the office of President."

Chapter 15

The Clash of Steel

Andrew Carnegie sold his steel company to J. P. Morgan in 1901, despite having misgivings about the way the deal was financed. What if Carnegie had refused to sell?

ON a cold night in December 1900, seventy-five of the nation's wealthiest businessmen had gathered at the New York University Club to honor Charles Schwab, president of the Carnegie Steel Company. When Schwab rose to speak, he began as he did every speech, saying that he would talk about steel because he wasn't able to talk about any other subject. But Schwab went on to describe a never-never land of the almost complete control of steel production, a dream in which the profits were unending and the competition almost nonexistent. He didn't have a name for this vision, but others knew it as vertical integration. Schwab was warning that his boss, Andrew Carnegie, was prepared to join a race with J. P. Morgan to become the first company to achieve full vertical integration of the steel industry.

It's not a concept that gets hearts beating in many people, but to a financial titan like J. P. Morgan, complete vertical integration of the steel industry was the full gospel, and

Schwab was giving an altar call. Morgan was too entranced by the speech to light his after-dinner cigar.

If a single corporation controlled the major steel company in the United States, Schwab said, and also controlled the iron ore mines and the factories that made finished products out of steel—steel pipe, barbed wire, tin plates, whatever—the entire enterprise would gain untold efficiencies. There would be lower transportation costs, both for goods and for deliveries, and fewer salesmen and managers would be needed.

Schwab left the most important element of his vision unsaid: As president of Carnegie Steel, creating such a consortium was exactly what he planned to do. This part of the dream would be a nightmare for J. P. Morgan. Morgan had bankrolled a new company, Federal Steel, and a collection of companies known as the "Americans": American Steel and Wire, American Tin Plate, American Sheet Steel, American Bridge, etc. He could put vertical integration into place almost immediately if he controlled his only major competitor, Schwab's company. But that presented quite a problem. As one wag later described the situation, "The cooks had discovered they had prepared and were ready to bake the finest plum pudding ever concocted, but that Mr. Carnegie had all the plums."

⤜∞⤛

In the late 1800s, American business was caught up in a frenzy of mergers. Across the country, small local businesses were selling out to large national trusts—from 1897 to 1904, one third of the companies in the United States were absorbed into larger companies. Consumers were not the beneficiaries of these mergers, however: Prices on goods increased as much as 400 percent after the competition was knocked out.

One holdout from the merger mania was Andrew Carnegie,

owner of Carnegie Steel, who was dubious of the trusts' stability. "They throw cats and dogs together and call them elephants," he said.

Carnegie didn't believe in buying out his competitors (with some exceptions). Instead he ruthlessly ran them out of business. He would find ways to produce his products cheaper and cut prices, and force the others to give up the game. Under Carnegie's reign as the king of steel, the price of steel fell from sixty-five dollars a ton to just twenty dollars a ton. Carnegie's customers were pleased, but his competitors weren't amused. This "unreasonable competition was childish and against public policy," one sniffed. Such child's play did work, though. In 1899 Carnegie Steel made a profit of $21 million; the next year, $40 million.

In response, J. P. Morgan decided to merge many of the small steel companies that were Carnegie's competitors into one large company, Federal Steel, which was headed by Elbert Gary (for whom the city of Gary, Indiana, was named).

Carnegie was dismissive of Federal Steel. "I think Federal is the greatest concern the world ever saw for manufacturing stock certificates . . . But they will fail badly in Steel," he predicted. Carnegie had reason to boast. His seven largest competitors produced 3.5 million tons of steel a year, while Carnegie Steel alone produced 3 million tons. It was the most dominant steel-producing company in the nation. Carnegie was contemptuous of companies that were overcapitalized through the issue of too many stock certificates, a practice known as financing with "water." Carnegie said that Federal Steel and another Morgan trust, American Steel and Wire, were financed "not merely with water but air."

These inflated companies became a threat to Carnegie Steel, however, when Morgan's companies such as American Steel and Wire cut their orders for steel from Carnegie and instead bought it from Federal Steel. But Carnegie knew how to deal with competition, and his philosophy was borrowed from the ruthless seventeenth century French chief minister Armand-Jean du Plessy de Richelieu: "First, all means to conciliate; failing that, all means to crush." Carnegie called together his board of managers and told them, "The situation is grave and interesting. . . . It is a question of the survival of the fittest."

Carnegie decided to expand his steel business into the markets that used his product. He began to build a $12 million rod mill on Lake Erie, which would compete directly with American Steel and Wire Company, an organization owned by one of Morgan's associates. Schwab had retained the rights to a new process of extruding tubes with hot steel instead of welding the tubes together. This new process would make more reliable tubes much more cheaply than the existing methods, and it was likely to drive American Steel and Wire out of business.

Carnegie announced that he would build a fleet of ships to carry the necessary materials on the Great Lakes, a move that would take business away from railroad magnate John D. Rockefeller. He added that he would not use Morgan's Pennsylvania Railroad to carry the product to the ports on the Atlantic Coast. He would build his own railroad. He also began buying land for new factories that would produce barbed wire, nails, tin plates, and other finished goods made of steel.

To the nation's business titans, Carnegie had become a dangerous madman. Morgan grumbled that "Carnegie is going to demoralize railroads just as he demoralized steel," but he knew that it would be he, and not Carnegie, who would be demoralized. Thus the most epic confrontation in American business was set in motion, a confrontation of business philosophies—competition versus combination. Just as companies and careers were about to be blown apart, into the divide came the skillful diplomat Charles Schwab.

Schwab was a smooth-talking salesman who, it was said, could talk the legs off of a brass pot. The thirty-eight-year-old manager had become a trusted second to Carnegie, and he was famous for his tact. It was against company rules for workers to smoke inside the steel mill, but if Schwab caught a worker with a cigarette, instead of firing him, Schwab would pull out one of his own expensive cigars and suggest to the worker that he put out his cigarette and enjoy the cigar, later, when he was off work.

After Schwab's speech at the University Club, J. P. Morgan asked to meet with him privately. The two men met secretly a few days later—a dangerous meeting for Schwab, because Carnegie would fire him if he knew that Schwab was meeting with the competition. At the meeting, Morgan and Schwab talked until three in the morning.

Morgan said that he wanted to stop Carnegie from expanding into fabricated products. Schwab insisted that Carnegie was committed to the new plans and that the only way to put an end to the old Scot's fight was to buy the entire company. Finally Morgan agreed that if Schwab could get Carnegie to name a price, he would try to meet it.

Schwab knew that Carnegie would resist giving up the empire that he had spent his entire life building. But Carnegie's wife was eager for Carnegie to leave the business, and for Schwab, there loomed the possibility that in this clash of steel titans, he was the one who would be crushed. Schwab waited for the right moment to present the idea, and he found it during a visit to the St. Andrew Golf Club in Yonkers.

Carnegie had taken up golf to improve his health (he referred to his outings as going to see "Dr. Golf"), and he was obsessed with the game even if he wasn't particularly skilled. Out on the course, Carnegie was pleasantly surprised that he beat the normally proficient Schwab. Carnegie was in a great mood when at lunch in his stone cottage on the course, Schwab laid out a vision for him, one of a life devoted to advancing knowledge and working for world peace, two passions of Carnegie's. Schwab then broke the news that Carnegie's rival, J. P. Morgan, was willing to buy the entire company.

The next morning, Schwab went to see Carnegie, and Carnegie handed him a scrap of paper on which he had written a price: $480 million dollars, in cash, bonds, and stock in the new company. Schwab took the scrap piece of paper to Morgan's office on Wall Street, handed it to him, and Morgan replied, "I accept this price."

On March 2, 1901, the biggest business deal of the twentieth century was finalized. For his part, Carnegie became the richest person in the country and, some said, in the world. Carnegie's bank had to build a special vault big enough to hold the $300 million worth of bonds that Carnegie had received in the deal.

Mr. Morgan now controlled the largest corporation in the world, the first to be worth more than $1 billion. U.S. Steel soon issued stock for $1.5 billion, although half of that was, to use Carnegie's term, air. When the first billion-dollar corporation was formed, a company's stock value was equivalent to its physical value (how much its factories and equipment are worth) plus one year's profits. Using this formula, U.S. Steel was worth approximately $750 million. Yet, Morgan issued stock valued at twice this amount—and this at a time when the entire U.S. stock market had a value of $9 billion.

Cosmopolitan magazine (a very different publication then from what it is today) wrote that "The world, on the 3rd day of March, 1901, ceased to be ruled by . . . statesmen. True, there were marionettes still figuring in Congress and as kings. But they were in place simply to carry out the orders of the world's real rulers—those who control the concentrated portion of the money supply."

William Jennings Bryan, the frequent presidential candidate who was not known for his humor, sarcastically wrote in the *Commoner*, " 'America is good enough for me,' remarked J. Pierpont Morgan a few days ago. Whenever he doesn't like it, he can give it back to us."

Finley Peter Dunne's commonsensical character Mr. Dooley threw in his two cents: "Pierpont Morgan calls in wan iv his office boys, th' prisidint iv a national bank, an' says he, 'James,' he says, "take some change out iv th' damper an' r-run out an' buy Europe f'r me,' he says. 'I intind to re-organize it an' put it on a paying basis.' "

Private investors could have been scared off by the high price of the U.S. Steel stock, but a feeling of irrational exuberance gripped the nation's investors, and instead of ignoring the stock and driving the price down, they rushed to buy it. People were gaga over the modern company and feared being left behind if they didn't own a piece of the action. Clerks, dressmakers, waiters, and other common workers poured money into the stock market, and the newspapers breathlessly reported how some of them had accumulated large amounts of money through their speculation. Barbers were said to earn thousands of dollars acting on tips they had received from the men they shaved, and in New York a slogan emerged on the street, "Buy A.O.T.," which meant "buy any old thing."

Everything was going up, and there was no way to lose. There were even cases of people being allowed to buy shares of U.S. Steel and other companies on margin, with generous terms that allowed them to purchase the stocks with 10 percent down and the remainder on monthly payments. Not surprisingly, a correction was soon in order—and it came in 1902— but a change had occurred. High finance was no longer for a few financial gods, but for everyman, for good or ill.

Other large mergers soon followed the example of U.S.

Steel. Automobile companies merged to form General Motors in 1908, motion-picture companies combined to form Paramount in 1916, and mining companies joined together to form Union Carbide and Carbon Company in 1917.

There would soon be efforts, led by President Theodore Roosevelt, to restrict and rein in these large monopolies and oligopolies, but such efforts only slowed the changes that were occurring. There were occasional market corrections. But for the most part, the market kept going up and up and up, buoyed by enthusiasm and water, until October 1929, when the dam burst and the fun was over.

WHAT IF?

What if Carnegie had decided not to sell J. P. Morgan his steel company?

In the stone golf cottage, Carnegie suddenly lost his appetite. Schwab had just proposed the sale of Carnegie Steel to J. P. Morgan. Carnegie studied Schwab's face, and Schwab turned his head to avoid meeting the Scot's eyes. Charlie had obviously been meeting with Morgan while he was supposed to be putting together the plan for the new plants on Lake Erie, a plan that would make Morgan never want to see anything made of steel again. Schwab's progress reports, were they all lies? Carnegie wondered. Suddenly Carnegie realized that Schwab had let him win out on the golf course that morning, and his face flushed with embarrassment.

"Charlie, do you remember what I told you at the board of managers meeting?" Carnegie said, still studying Schwab's face for a clue of what was really happening. "Do you remember that I told you it was going to be survival of the fittest? Do you think that Carnegie Steel is less fit than that collection of would-bes and has-beens that Pierpont has assembled? I tell you, Charlie—I've said it before—you can throw any number of cats and dogs into a sack and you'll never end up with an elephant.

"I'll tell you another thing," Carnegie continued, his face red above his white beard. "My father used to sing an old ballad to me, 'Wha daur meddle wi' me?' Pierpont will always regret that he decided to meddle with me."

Carnegie had been thinking about retiring, his wife was hoping that he would, but with the challenge set down before him by J. P. Morgan and his army on Wall Street, Carnegie threw himself into his steel business with an enthusiasm that he hadn't had for ten years. He pushed and threatened and cajoled the workers building the factories until the first of the new factories, the one that was making steel pipe and tubing, was ready to begin production just a year later, in 1902. Carnegie was able to produce the pipe for ten dollars less a ton than American Steel and Wire, and to increase the pressure, he began selling his superior product at cost. Within another year, American Steel and Wire was forced to drop out of the steel pipe business and concentrate on barbed wire, its managers watching fearfully as Carnegie's wire factory neared completion.

American Steel and Wire wasn't the only company in trouble. Carnegie was putting pressure on all of the companies in Morgan's constellation. Carnegie had been right about the companies: They were woefully overcapitalized by their aggressive sale of stock—they were completely waterlogged. As the companies were subjected to Carnegie's intense competition, one by one they faced a panic by their investors, and they began to sink from sight. Morgan was also having difficulty raising new money from Wall Street because Carnegie, by his actions and his frequent speeches, was able to puncture almost all confidence in the stock market as a stable investment. By 1908, only Federal Steel was left of Morgan's former empire.

In the late 1890s, Carnegie was making an annual income of $15 million, several thousand dollars more a day than the president was making each year. At the time, the average American laborer drew an annual salary of under five hundred dollars, three hundred dollars in the Southern states. Nearly one out of five children worked in factories, no doubt some of them in factories owned by Carnegie, Morgan, Gary, et al. By 1908, Carnegie was making ten times what he had been making a decade before, while the incomes of the common laborer had not increased significantly over the same period.

There were threats and rumors of massive strikes, as had happened in the coal mines, but Carnegie acted quickly to head off any major interruptions at his companies. He didn't raise wages—instead, in every city and town where Carnegie workers lived, Carnegie donated money to build a new public library, and to each town's Presbyterian church he offered a new pipe organ. These things were expensive, but Carnegie knew that they were millions of dollars cheaper than increasing wages for all of his thousands of employees.

Carnegie had been correct that the nation's industries could produce a stable economy, but workers continued to live in poverty while a few barons of industry lived lives of unimaginable extravagance. The nation's economy was stable, unless someone tipped over the monopoly board, and that someone was soon to appear.

In 1901 labor organizer Eugene V. Debs had founded the Socialist Party of the United States. He had guided the Pullman strike of 1895 and been thrown in jail for his efforts, but while in jail he had done much reading and came to hate the excesses of capitalism. Debs had run for president on a populist platform in 1900 and as a Socialist in 1904 and 1908, and had not been taken seriously in any of the elections. But in the election of 1912, when Teddy Roosevelt formed the Bull Moose party and split votes away from Republican candidate William Taft and Democrat Woodrow Wilson, Debs cobbled together a collection of voters who deplored the disparity of personal income between the owners and the factory workers. Debs's supporters saw the disparity as undemocratic, and Debs was elected president.

Carnegie, Morgan, Cornelius Vanderbilt, John D. Rockefeller, and the other pillars of U.S. industry viewed Washington politicians as minor irritants to their daily operations. They were concerned about Debs's election. All of the business titans considered him a rabble-rouser who would make their lives harder by stirring up more ill feelings in their workers. But they were completely unprepared for what Debs truly had in mind.

In 1914 Carnegie finally forced the last of Morgan's companies, Federal Steel, to a state of near oblivion, with its formerly heavily watered stock trading for less than one dollar a share. But Carnegie and his fellow financial barons were surprised when the company was sold to, of all things, the United States government.

Citing national security and financial stability, President Debs bought up Federal Steel and several other prominent companies, thereby nationalizing a significant portion of the nation's industry. Debs then enacted what he called a "New Square Deal" (echoing Teddy Roosevelt's reform program called the "Square Deal"), and pushed through new high taxes on corporate profits and personal income. At the highest levels, corporate profits could be taxed at 90 percent.

Carnegie had predicted that the financial landscape of America would be forever changed by the collapse of the stock exchange. But it was Socialism that finally ended the reign of the Wall Street bankers in America.

Carnegie sold his company, but he came close to having it handed back to him. He had refused stocks and insisted on bonds instead, and when the waterlogged U.S. Steel almost defaulted on the bonds a few years later, Carnegie nearly assumed ownership of the entire operation.

For Carnegie it was probably a relief that U.S. Steel was able to recover, because his days as a ruthless capitalist were behind him. He was now Andrew Carnegie, world-famous philanthropist, and he enjoyed this role much more than anything else he had done. Carnegie had said that a man who dies with millions of dollars sitting in the bank would die "unwept, unhonored, and unsung." "The man who dies thus rich," he said, "dies disgraced." Carnegie quickly went to work at giving away his money. He helped the poor, he built 2,811 free public libraries around the world, he gave money to launch what is now Carnegie-Mellon University, he donated almost eight thousand organs to churches, and he set out to speak on world peace. (He was bitterly disappointed at the outbreak of World War I.)

J. P. Morgan went on to dominate not just steel production but much of American business. Although the stock market rose and fell with alarming frequency and intensity in the decades following the deal between Carnegie and Morgan, Morgan was able to fly above such fiscal concerns. In 1931, he could truthfully say, "I don't know anything about a depression."

Chapter 16

Teddy Roosevelt Outlaws Football

❧

What if Teddy Roosevelt had followed through on his threat to abolish the nation's most popular collegiate sport?

If U.S. presidents were winds that directed the nation one way or another, Teddy Roosevelt (TR), the nation's twenty-sixth president, would have been a tornado. He was a man of voluminous energy and mountainous intellect, and to the horror of some and the delight of many, he applied his talents to every issue that attracted his attention. While president, Roosevelt reformed the railroad trusts, and he pushed through reforms of the food, meat-packing, and drug manufacturing industries. To solve the problem of the long voyage around South America, TR initiated work on a canal through Central America—and when the nation that owned the land, Colombia, insisted on payment for the trespass, he helped to create a new nation—Panama—that was more agreeable to his desires. An avid outdoorsman, TR created the national park system. He also won the 1905 Nobel Peace Prize for negotiating an end to a war between Russia and Japan.

Then there were his lesser interests: Roosevelt tried to have "In God We Trust" removed from U.S. coins because he thought that mixing God and mammon was vulgar. Then he launched a campaign to simplify spelling, a proposal that prompted the *Louisville Courier-Journal* to mock, "No subject is tu hi fr him to takl, nor to lo for him to notis."

In 1905, Roosevelt found another subject to preach about from his bully pulpit: collegiate athletics. "[Roosevelt] today took up another question of vital interest to the American people," grumbled the *New York Times* in 1905. "He started a campaign of reform of football."

⁂

In the late 1890s, football was the latest craze on college campuses, and every college and university in America tried to field a team. Football had its origins in rugby, but the new sport added yard marks, introduced the idea of downs, and allowed the team to retain possession after the ball carrier was tackled. Although football in the late nineteenth century resembled the rough game of today, the early version was much more dangerous. At the time, the rules allowed players to begin running toward the line before the snap of the ball. Once the ball carrier hit the line, players on both teams would pull, tug, crash, and punch to move the pile of players in the proper direction. It was in these pileups that many young men were severely injured The rough spirit of the sport meant that beatings in the pile were a part of the game, and these beatings were often quite severe; in a game on Thanksgiving 1894, for example, a Georgetown University player was killed.

Perhaps more alarming to university faculty and administrators were the outrageous ethical gymnastics many colleges were willing to go through to win on the football field. The schools often let players cheat on their college entrance exams, and rumors abounded that many schools were paying the best players. When one Penn State athlete had an especially good game against the University of Pennsylvania, he showed up the very next week practicing with the Penn team. The following year he played for both Penn State and Penn. At Yale, a student was given the entire cigarette commission for the campus in exchange for being on the team. In the Midwest, it was discovered that seven of the University of Michigan's eleven starting players weren't actually enrolled at the University. At Purdue University, after opposing teams suspected the university of employing railroad toughs to play on its team in place of college students, opponents began derisively referring to the team as the Boilermakers, a nickname that stuck.

As a result of such abuses, in 1894, Charles Eliot, the president of Harvard University, decided to lead a crusade against college athletics in general and football in particular. Football, he complained, had put universities into the entertainment business. Instead of improving character, Eliot said, football was responsible for spoiling young men and turning them into "powerful animals." At the University of Wisconsin, historian Frederick Jackson Turner agreed, saying that it was an "absurd idea" that college athletics was "the test of the excellence of a university and the proper means of advertising it." (When Turner proposed suspending football at Wisconsin for two years, a group of students gathered at his house determined to throw the good professor into Lake Mendota.)

Soon muckraking journalists took up the story, and articles began appearing in several publications with examples of the violence and ethical lapses behavior occurring on campus athletic fields. *Collier's* magazine reported on how Big Ten universities tried to purchase victories on the gridiron, and the *Nation* wrote articles calling for the complete abolition of football. But the most important article appeared in *McClure's Magazine*, not just for what it said, but because the journalist, Henry Beach Needham, was a friend of President Roosevelt. The article asserted that football had become a critical business for many of the nation's universities and showed that at Harvard, football receipts accounted for $76,000 of the $112,000 brought in by all twenty-four sports.

Roosevelt himself was a big fan of the game; in fact, his son played on the freshman team at Harvard. One year, after Yale triumphed over Roosevelt's beloved Crimson, one of the president's cabinet officers make a crack that Harvard should replace Yale with Vassar on the schedule, and Roosevelt nearly lost his composure. "I behaved with what dignity I could under distressing conditions," he later said of the incident.

When it came to social reform, Roosevelt wanted, as a friend of his once said, "reformers who ate roast beef." Roosevelt wanted the universities to reform football, but just the excesses. The sport itself, he found, was of value. In the fall of 1905, Roosevelt gave a speech at Harvard on the subject:

> *I believe in outdoor games, and I do not mind in the least that they are rough games, or that those who take part in them are occasionally injured. I have no sympathy whatever with the overwrought sentimentality which would keep a young man in cotton-wool, and I have a hearty contempt for him if he counts a broken arm or collar-bone as of serious consequence when balanced against the chance of showing that he possesses hardihood, physical address, and courage . . .*

However, Roosevelt, said, he did think reforms were needed in the way the universities were treating the athletes:

> *It is a bad thing for any college man to grow to regard sport as the serious business of life . . . The college undergraduate who, in furtive fashion, becomes a semiprofessional is an unmitigated curse, and that not alone to university life and to the cause of amateur sport; for the college graduate ought in after-years to take the lead in putting the business morality of this country on a proper plane, and he cannot do it if in his own college career his code of conduct has been warped and twisted. . . . It is a better thing for our colleges to have the average student interested in some form of athletics than to have them all gather in a mass to see other people do their athletics for them.*

That fall the collegiate football season was more deadly than ever—twenty-three players died from injuries suffered on the field. When President Roosevelt saw a newspaper photograph of one of these players, a bloodied and badly beaten player from Swarthmore, he threatened to abolish the game nationwide.

(There appears to have been more bluster than usual in Roosevelt's threat—he would later write to a friend that his alma mater of Harvard would be "doing the baby act" if they followed university president Eliot's suggestion of unilaterally abolishing football in Cambridge.)

Following the 1905 season, a group of university presidents met at the invitation of the president of Columbia University at the Murray Hill Hotel in New York City. At the meeting, the chancellor of New York University compared the athletic departments to the Russian monarchy, saying that in that country, "it is the Russian people against the Russian grand dukes. Here it is the football people against the football grand dukes." As expected, a resolution was introduced at the conference, stating: "That the game of football as played under existing rules shall be abolished."

Few in the country took the concerns of handwringing university administrators seriously, except for Bill Reid, the twenty-six-year-old coach at Harvard. Reid knew that the football establishment, which was dominated by Yale's coach, Walter Camp, would never agree to major changes in the game. But he was worried that if football was abolished, as President Roosevelt was suggesting, he would naturally lose his job. It was a good job for a young man, too—as football coach he made a higher salary than any Harvard professor, and his salary was almost as high as President Eliot's. Reid decided to publicly propose radical rule changes, saying that if the rules weren't adopted, "there will be no more football at Harvard, and if Harvard throws out the game, many other colleges will follow Harvard's lead, and an important blow will be dealt to the game."

Eliot didn't trust Reid and the other coaches to clean up the game: "It is childish to suppose that the athletic authorities which permitted football to become a brutal, cheating, demoralizing game can be trusted to reform it," he said. But at Reid's prodding, many changes were adopted. There was opposition from the established coaches, who formed what as known as the "athletocracy." Amos Alonzo Stagg, coach at the University of Chicago, complained that the new rules had turned the sport into a "parlor game," and Yale's Walter Camp complained that Reid's new rules were just "a big Bluff" and that they were "so radical that they would practically make a new game."

But the universities agreed to the new rules. Penalties were added for violence; the forward pass was introduced; six men had to begin on the line of scrimmage, with only a ball-length neutral zone between them (which prevented the flying-wedge formation from gaining steam behind the lines); and the teams were required to move the ball ten yards in three downs. Camp had been correct in many aspects: It was a new game, but one with an emphasis on moving the football, not on punches and thuggery.

The very next October, the new era of football became a reality when the quarterback for Connecticut's Wesleyan University, Samuel F. B. Moore Jr., threw a twenty-yard pass to his halfback, and many other colleges quickly adopted the pass play, which relied more on finesse than brute strength. The game was still a rough contest, and other young men would die or become seriously injured playing it in the years to come. But because of Roosevelt's public criticism and the resulting reforms, football continued as the most popular sport in America.

WHAT IF?

What if President Theodore Roosevelt had persisted with the plan to abolish football on the nation's college campuses?

In January of 1906, the president of the United States asked for the presidents of several of the major universities to attend a meeting at the White House. The leaders of Harvard, Yale, Princeton, Columbia, Georgetown, Virginia, Wisconsin, and Michigan dutifully made the long train trip to Washington like schoolboys called into the principal's office for roughhousing on the playground. (A few did grumble that it took a dispute over football for them to finally receive an invitation to the White House.)

In Roosevelt's office, the president introduced Charles Eliot, the president of Harvard, who was to give the introductory remarks for the meeting. "From the president of the United States to the humblest member of our faculty . . ."—Eliot nervously paused as a few of the men chuckled at the idea of humble faculty—"there is universal condemnation of this boy-killing, education-prostituting, mammon-making, gladiatorial spectacle that . . ." Eliot droned on for twenty minutes, listing the many offenses that the game of football had committed on the nation's college campuses. Roosevelt didn't want his guests to begin falling asleep as Eliot went beyond the prelude to his soliloquy, so he stood up and moved around his desk and began pacing in front of the lectern set up for Eliot's opening remarks. Finally he couldn't hold his tongue any longer, and he interrupted the Harvard president.

"You gentlemen may already know, my own son, Ted, is on the freshman team at Harvard. Some of you may have seen the newspaper stories about the game with Yale, when his arm was broken. I was concerned that he had been singled out because of my comments about football, but when he told me that his arm had been broken fair and square, I told him that such injuries would only serve to improve his hardihood. For that is what I believe. We should not mollycoddle these young men.

"However, our nation cannot stand to see"—Roosevelt paused, and one of his aides passed him his reading glasses as he strained with his good eye to read a piece of paper—"to see twenty-three of its finest young men killed over a sporting activity. Have you men seen the types of things that are happening on the football field? Look here." Roosevelt passed around the photograph of the Swarthmore player that had disgusted him so. "If you gentlemen have attended the football matches, you have no doubt seen the types of cheating that go on in the name of sport. Look here. President Eliot, come at me as if you were a footballer with the ball."

Eliot had remained standing at the lectern, not sure whether he was expected to return to his seat or continue his opening remarks when Roosevelt had finished. Now he looked at the other men, who were all smiling, as the president removed his glasses, crouched down in a football stance, and gestured with his hands that Eliot should charge him. Eliot gamely began shuffling awkwardly toward Roosevelt, when TR rushed forward and grabbed him in a gentle half-tackle, holding Eliot slightly off balance with his left foot off the floor. "Now Ted has shown me that when a man is being tackled and unable to protect himself, often the defender will take his fist and strike here," Roosevelt said, pantomiming a punch to the Adam's apple, "or even here," TR continued, imitating a punch to the groin.

"Then the other defenders fall on the ball carrier and pummel him," Roosevelt said, setting Eliot back on both feet. "This is not the way to train our nation's best young men for the nation's affairs. We cannot stand by and let our nation's citadels of learning become theaters of brutality.

"If you gentlemen will ban the game from your universities, I am certain that other activities can be found for your students that will improve their hardihood without corrupting their minds."

Each one of the of university presidents, fearing that Roosevelt might call them forward to demonstrate other football moves that he had learned, promised the president that the 1905 season would be the last that football would be played on their campuses.

At the universities and colleges, football and other intercollegiate sporting events were discontinued in favor of mandatory physical education classes for all students. The athletes on the football teams were surprised to learn in the spring of 1906 that President Roosevelt had "outlawed" football, as it was explained to them by their coaches. It didn't occur to one of them that such an act would be unconstitutional. Almost immediately, nearly every one of these student athletes shared an epiphany that their academic careers mattered very little to them. The next autumn, most of the nation's collegiate footballers failed to enroll in classes.

Although some students were frustrated by rules that required chemistry majors to be able to swim the length of a pool in order to graduate, university administrators heard few complaints. The sporting public soon looked elsewhere for its entertainment.

Football continued even without the academic endorsement. The game quickly evolved into a more modern action-oriented sport with less thuggery. In 1920, a group of men met in Ray Hay's auto showroom. Sitting on running boards, the men launched the American Professional Football Association, which changed its name to the National Football League two years later. Needing to establish a way to develop young players, the league allowed smaller cities to join as minor league development teams. By the 1950s, the annual Federal Football League championship was one of the biggest sporting events of the nation. Although the Federal championship was for the best of the junior professional league, because more than seventy-five cities had Federal Football League teams, there was broad national interest in the outcome.

With no sporting events to hold their attention, Americans had much less interest in the happenings on the nation's campuses. In the 1920s, most Americans didn't go to high school, and a college education was perceived as something reserved for the wealthy and for a few, rare academic marvels. The universities' isolation from mainstream America would soon become disastrous for many of the universities, however. By 1932, in the middle of the Great Depression, nearly one out of four Americans was out of work, and most states were faced with plummeting state tax revenues. As a response, many states eliminated or dramatically cut back their funding for state universities. In 1929 there had been nearly three hundred state-supported colleges and universities in the nation; by 1939 there were fewer than seventy-five. Many of the schools that remained were more properly described as "state-assisted" instead of "state-supported," and the institutions were forced to raise their tutitions almost as high as those of the private institutions to compensate. As a result, only one person out of a hundred could afford to attend a state university.

Roosevelt's reforms helped to make a popular game even more so. The football coaches who met to rework the rules of the game continued their association, which eventually evolved into the National Collegiate Athletic Association, or NCAA.

Moreover, as a result of college athletics in general and football in particular, the well-being of colleges and universities became the concern of masses of people who had never attended any of the institutions. As one wag wrote at the turn of the century, sporting events have as much relevance to education as "bullfighting has with agriculture." But if it weren't for the brutal sport of football, the refined scholarly work that takes place in the nation's universities might never have flourished, let alone been available to countless young American men and women.

Chapter 17

The Feast of *The Jungle*

Upton Sinclair's novel *The Jungle* brought about changes in the nation's food system. What if he had listened to those who said that his novel wasn't publishable?

AT the turn of the twentieth century, a British traveler to the Midwest wrote about Chicago, "Great as the city is in everything, it seems that the first place among its strong points must be given to the celerity and comprehensiveness of the Chicago style of killing hogs." But although Carl Sandberg later tagged the city "hog butcher for the world," it was beef that made Chicago wealthy.

Beef, unlike pork, could not be easily preserved. Pork can be eaten months after butchering as ham, bacon, or other smoked or salted meats. But although some New Englanders developed a taste for pickled beef, for the most part people preferred to eat their steaks fresh.

Until the 1800s, this meant that cuts of beef were prepared by local butcher shops from local livestock. But in 1872 a gentleman named Gustavus Swift began buying cheap cattle from the Western ranges and shipping sides of beef from

Chicago to the cities on the East Coast on express trains that ran in winter with their doors open to preserve the meat. Later Swift developed refrigerated railroad cars, and soon he and his competitor, Philip Armour, were delivering most of the nation's meat.

The slaughterhouses were located on the south side of Chicago in an area known as Packingtown, a slum populated by the Eastern European immigrants who worked in the packing plants. The plants were dirty, disgusting workplaces that treated the workers with little more regard than was shown the livestock that passed through the doors.

The owners of the packing plants liked to brag that they used every part of the hog but the squeal, but there were still scraps to be disposed of, and the slaughter refuse was dumped into a large open sewer that ran down the middle of Packingtown called "Bubbly Creek," so named because it would literally bubble from the microorganisms emitting gas as they went about their business of devouring the refuse.

The horrible conditions of Packingtown caught the attention of Frank Warren, the editor of a weekly Socialist newspaper, *Appeal to Reason*, and Warren decided to send one of his frequent contributors, a twenty-six-year-old man named Upton Sinclair, to Chicago to write an exposé.

Sinclair was destitute himself, and so his clothes were as threadbare and torn as those of the Eastern European immigrants who worked in the meat-packing plants. Sinclair found that by simply carrying a pail, he could go anywhere in the plant that he wanted. He was able to gather his material, and the series began running in *Appeal to Reason* in February 1905. Almost immediately Sinclair began trying to convince book publishers to use the articles as the basis for a novel.

Five publishers rejected the book, including Macmillan, the publisher of Sinclair's previous book. Sinclair received little encouragement for the novel. When he invited Warren to his home to read to him the new ending of *The Jungle*, Warren fell asleep. And when Sinclair went to see the famous muckraking journalist Lincoln Steffens about the publishers' lack of interest, Steffens advised Sinclair to give up: "It is useless to tell things that are incredible, even though they may be true," Steffens said.

Finally, the sixth publishing company that Sinclair sent the manuscript to, Doubleday, was interested but had concerns about the accuracy of the book. One of the editors sent a copy of the manuscript to the managing editor of the *Chicago Tribune*, James Keeley, asking about its veracity. Keeley sent back a thirty-two-page report, which he said had been prepared by one of his reporters, that asserted that the book was a sham, a complete fiction. The world of foul working conditions and unhealthy meat that Sinclair had described simply didn't exist, the report said. Sinclair desperately protested that the book was accurate and convinced his editor at Doubleday to send someone else to Chicago to discover the truth. The editor sent an attorney, who found not only that Sinclair's accusations against the meat-packing industry were accurate, but also that the "reporter" that the managing editor of the *Chicago Tribune* had sent out to investigate was actually a public relations flack employed by the meatpackers.

When *The Jungle* was finally published in February of 1906, the book made headlines across the nation—except in Chicago, where the newspapers ignored it. The novel focuses on the family of a Lithuanian immigrant, Jurgis Rudkus, who works and lives in Packingtown. The book attempts both to manipulate readers' emotions, with the death of Jurgis's little son, and to shock, with scenes such as the one in which a worker falls into a rendering vat and becomes a part of the lard. The novel was best known, however, for its graphic detail:

There was never the least attention paid to what was cut up
for sausage; there would come all the way back from Europe
old sausage that had been rejected, and that was mouldy and
white—it would be doused with borax and glycerin, and
dumped into the hoppers, and made over again for home
consumption. There would be meat that had tumbled out on
the floor, in the dirt and sawdust, where the workers had
tramped and spit uncounted billions of consumption [tuber-
culosis] germs. There would be meat stored in great piles
and thousands of rats would race about on it. It was too
dark in these storage places to see well, but a man could run
his hand over these piles of meat and sweep off handfuls of
the dried dung of rats. These rats were nuisances, and the
packers would put poisoned bread out for them; they would
die, and the rats, bread and meat would go into the hoppers
together.

By the end of the year, Sinclair's book had sold more than
one hundred thousand copies, and an estimated one million
people had read it. Sinclair became so well known that the *New
York Evening World* reported, "Not since Byron awoke one
morning to find himself famous has there been such an example of world-wide celebrity won in a day by a book as has come
to Upton Sinclair." Jack London, a writer who encouraged
Sinclair's Socialist writings, praised the book as "the *Uncle
Tom's Cabin* of wage-slavery." In England the book was favorably reviewed in a two-part series by the thirty-two-year-old
journalist Winston Churchill. But it was a reader in a special
position who gave the book its place in history.

"I have now read, if not all, yet a good deal of your book,"
President Theodore Roosevelt wrote to Sinclair. "If you can
come down here . . . I shall be particularly glad to see you."
Roosevelt himself was considered a reformer, but he held

muckrakers such as Sinclair at arm's length. (He had even coined the pejorative term "muckraker" from a line in John Bunyan's *The Pilgrim's Progress:* "A man that could look no way but downwards with a muckrake in his hand.") But Roosevelt had criticized the nation's meatpackers for years—during the Spanish-American War, Roosevelt said that he would as soon eat his old hat as eat the meat the companies were sending to Cuba—and he thought that Sinclair's book might help him do something about the problem.

The humorist Finley Peter Dunne, in the guise of his character Mr. Dooley, speculated on Roosevelt's reaction to the book: "Tiddy was toying with a light breakfast an' idly turnin' over th' pages iv th' new book with both hands. Suddnnly he rose fr'm th' table, an' cryin': 'I'm pizened,' began throwin' sausages out iv th' window . . . Since thin th' Prisident, like th' rest iv us, has become a viggytaryan. . . ."

Roosevelt began actively pursuing government regulations on the meat industry, including regular inspections. The meatpackers, not surprisingly, were opposed to the regulations and new scrutiny. Such measures, they said, "will put our business in the hands of theorists, chemists, sociologists, etc. and take management and control away from men who devoted their lives to the upbuilding of this great American industry."

Soon, with meat sales falling, the meatpackers and their friends couldn't stand by quietly any longer. An article ghost-written for Ogden Armour that appeared in *The Saturday Evening Post* said, "Not one atom of any condemned animal or carcass finds it way, directly or indirectly, from any source, into any food product or food ingredient." The *Chicago Tribune* finally mentioned *The Jungle* in a dismissive editorial: "The conditions depicted in *The Jungle* . . . are the product of a distempered imagination and credulous mind of a pseudo social reformer."

But many people believed *The Jungle*. The book inspired a popular rhyme in 1906, "Mary had a little lamb, and when she saw it sicken, she shipped it off to Packingtown, and now it's labeled chicken."

Then, in the midst of the debate, physicians began supporting Sinclair's claims. Caroline Hedger, M.D., who was working in the packinghouse area of Chicago, published an article saying that the Packingtown district had the highest rate of tuberculosis of any area in the nation. She claimed that this high rate of infection was caused by the unsanitary conditions in the packing plants, and that the disease was being transmitted to the general public through bacteria-laden meats. Another Chicago physician, W. K. Jaques, M.D., wrote of his own inspections of the meat plants, which he called the nation's kitchens. "The beef trimmers cut off unsightly portions, bruised or injured places, enlarged glands or abscesses. I asked the [city] inspector what was done with these trimmings. 'Sausage,' was his laconic reply." Jaques, who noted that the city health inspector was responsible for preventing such abuses, suggested, "It ought not ever be possible that the quality of the nation's meat should depend on the conscience of a Chicago politician."

Meanwhile, President Roosevelt had commissioned a report on the packing plants, and he waited until meat inspection legislation was being debated in Congress to release the results. The report confirmed Sinclair's accusations and also supported the claims that the meat was spreading disease throughout the nation. Although the powerful Speaker of the House, "Uncle Joe" Cannon of the packers' home state of Illinois, was able to water down the new regulations, less than a year after the publication of *The Jungle*, the nation's first meat inspection bill was passed. Sinclair's book also ushered in a new era of governmental oversight of the nation's health, not only with meat inspections, but also other foods and drugs, resulting a few years later in the creation of the Food and Drug Administration.

WHAT IF?

What if Upton Sinclair had decided five rejection letters were enough and hadn't tried again to publish his novel?

Lincoln Steffens was right. That was the best explanation for his failure to find a publisher for his book, Upton Sinclair decided. What person, except someone living in Packingtown, would ever believe how the workers there were treated or how the meat there was prepared? Perhaps more important, what person would ever want to read about it?

"The prospects for getting my book published might be even worse than I realized," a despondent Sinclair wrote to his friend Jack London. "Who knows what lies these companies are telling the publishers? I would consider publishing it myself, but as it is now, I have so little money that it is a struggle to buy food to eat." Sinclair realized that he was just another undercapitalized Socialist.

The next year he tried to grow his own food in a garden, with disastrous results. Sinclair was a poor farmer, and even this meager attempt at self-sufficiency was a failure. Over the next winter, his life became increasingly difficult and in a few instances he seemed to lose his grip on reality. Finally, in the summer of 1907 Sinclair's family decided to send him to a mental institution in Michigan. At the Battle Creek Sanitarium, Sinclair met John Kellogg, who was treating patients by varying their diets to include more nuts and dried vegetables. Sinclair eagerly joined Kellogg's experiments and in fact developed several new cereal products himself. Kellogg's brother, W. K. Kellogg, had started a breakfast cereal company to market John's new healthy foods, but the Kellogg brothers soon found themselves in competition with two of their former patients, Sinclair and C. W. Post, each of whom also launched cereal companies. The morning convenience food was popular enough that all three companies, Kellogg, Post, and Sinclair, were successful.

The meatpacking industry also continued to thrive. After Sinclair's articles appeared in *Appeal to Reason* in 1905, the executives of the companies knew that some type of response was needed. If the public believed that their meats were dirty, impure, or even infectious, sales would disappear. Sinclair had forced their hand—they had to make major changes in the way that they conducted their business.

Within weeks after Sinclair's articles in *Appeal to Reason*, each of the major meatpacking companies immediately doubled their public relations staffs, put more attorneys on retainer, and increased the number of security guards at their plants. Thus protected, they were inoculated against the next "reformer" who tried to expose what was happening in the packinghouses of Chicago.

These changes did nothing to improve the quality of the meat that was being shipped across the country. The products coming out of the factories, especially the prepared meats such as bologna and hot dogs, were menageries of infectious bacteria. *Escherichia coli, Listeria monocytogenes, Mycobacterium tuberculosis,* and several types of salmonella, including *S. typhimurium, S. choleraesuis,* and *S. typhi,* all flourished and frolicked in the perfect conditions created for them by the meatpackers. Because the diseases caused by such bacteria were common throughout mankind's history, contemporary outbreaks brought little notice. It was common for people to become sick with the "stomach flu" after eating at a local restaurant, or to be a bit nauseated after stuffing themselves at a cookout. Occasionally a more serious disease, such as tuberculosis, appeared, but no one considered food as a possible culprit. When foods were identified by determined health officials as the cause of an outbreak of illness, the story was generally considered to be a local one, and few people across the nation were aware of any type of problem.

All of that changed in the 1950s. In 1955, a restaurant-supplies salesman named Roy Block noticed that one hamburger stand in Bakersfield, California, was buying an enormous number of deep fryers designed to cook french fries. Block traveled to see the hamburger stand and discovered a miracle of feeding efficiency. The hamburger stand, which was owned by the three Callahan brothers, had created a system that was ideal for producing fast food. The combination of labor specialization and task routinization produced food of standard quality, and it also had the not insignificant benefit of allowing the restaurant to be run by an inexpensive teenage work force. In 1955, Block asked the Callahan brothers if he could be the sole agent franchising others for the hamburger stand, and he built his first Callahan's drive-in outside of Phoenix, Arizona.

Capitalizing on the Daniel Boone-Davy Crockett–inspired frontier fad that was popular with children at the time, the Callahan restaurants were designed to look like frontier forts, except instead of logs the forts looked as if they were constructed of giant fiberglass french fries. To attract even more children to his restaurants, Block hired a local weather announcer in Chicago to dress up in a fringed-leather costume and a coonskin cap, and presented him as "Dan Callahan, the roughest, toughest hamburger-eatin' man on the frontier." Soon Callahan french fry forts were appearing at every major intersection in suburban America, quickly followed by me-too franchises also selling the dietary staples of burgers, fries, and colas.

In 1962, the Callahan restaurants experienced a slump in sales caused, marketing surveys showed, by many people's dislike for the taste of the overcooked, cracker-thin Callahan burgers. The hamburgers, customers complained, didn't taste as good as hamburgers made at home. The company responded in the fall of 1965 with the "Big Dan," a hamburger that was three times the thickness of an ordinary burger, with more than a quarter-pound of beef even after it was cooked. "Three times the burger for three times the man, that's Big Dan!" jingled the television commercials.

But the Big Dan turned out to be big trouble. The larger burger didn't always cook through on the grill as well as the smaller burgers had. In July 1966, a Callahan's food processing plant had allowed a shipment of meat to wait on the loading dock too long, and the ground beef, which was already contaminated with bacteria, soon developed alarming levels of pathogens. In keeping with the company's typical efficiency, the ground beef was formed into thousands of identical Big Dan patties, which were then shipped to seventeen states in the Midwest and West. Within days, reports of outbreaks of illnesses began, first an elderly person in Madison, Wisconsin, then a four-year-old boy in Des Plaines, Iowa, then two teenagers in Columbus, Ohio. Within a week, more than forty-seven people had died from the nation's worst food-borne illness outbreak.

Consumer activist Ralph Nader soon released a report on the nation's meatpacking industry. Titled *Unsafe at Any Heat*, the report exposed the unhealthy conditions at the beef- and pork-packing factories in the Midwest. Nader's report pointed out the enormous number of food-illness outbreaks in the nation and prompted meat inspection legislation that led to major changes in the way meat and poultry were prepared for consumption in the United States. But it was too late for Big Dan Callahan, who would brave the chrome-bumpered frontier no more. The public no longer trusted the quick and cheap fast-food restaurants, and they returned to local diners and coffee shops for their noonday meals. Although other fast-food restaurants would later thrive, the original fast-food hamburger joint, Callahan's, went bankrupt, done in by the poor quality of the nation's meat supply.

Despite the fact that Sinclair's book had allowed Theodore Roosevelt to institute what he saw as needed reforms to improve the nation's health, Roosevelt never warmed to the earnest Sinclair. After their initial meeting, Sinclair began bombarding Roosevelt with letters, magazine articles, and other information extolling Socialism. Roosevelt soon tired of Sinclair's pestering and sent a letter to Frank Doubleday, the publisher of *The Jungle:* "Tell Sinclair to go home and let me run the country for a while."

Roosevelt later wrote, ". . . in the meat-packing business I found Sinclair was of real use. [However] I have utter contempt for him. Three-fourths of the things he said were absolute falsehoods. For some of the remainder there was only a basis of truth. Nevertheless, in this particular crisis he was of service to us . . . I could not afford to disregard ugly things that had been found out simply because I did not like the man. . . ."

For their part, the meatpackers, who had so strenuously opposed the intervention in their business just months before, soon turned the federal inspections into a marketing opportunity. They began advertising that their meat products were guaranteed pure by the U.S. government.

Sinclair never had another journalistic success that approached that of *The Jungle*, and he was never able to bring about the types of social reforms that he dreamed of. Because of this, he was disappointmented by the response the book provoked. Sinclair himself best summed up the turn of events: "I aimed at the public's heart, and by accident, I hit it in the stomach."

Chapter 18

Wilson Creates Peace, Not War

What if President Woodrow Wilson had not used the Zimmermann telegram to initiate America's entry into World War I?

IN the early part of the twentieth century, the nations of Europe attempted to avoid war by forming treaties tying one to another, like mountain climbers scaling a peak linked by a rope. What these nations hadn't counted on was that when one fell, they would all go over the precipice.

On June 18, 1914, the man who would be king of the Austro-Hungarian empire, the Archduke Francis Ferdinand, and his wife were both shot and killed by a Serbian terrorist in the Bosnian city of Sarajevo. Leaders of Austria-Hungary accused the government of Serbia of the murder. After a month of rising tempers, on July 29, 1914, Austria-Hungary declared war on Serbia and began shelling the capital city of Belgrade. Russia immediately declared war on Austria-Hungary over the bombing of Belgrade, and then on August 1, Germany declared war on Russia because of Germany's alliance with Austria-Hungary. France, which had formed an alliance with Russia, then declared war on Germany. Britain had decided to stay out of the conflict, but when the German army decided to attack France by marching through Belgium, Britain declared war over the violation of Belgium's neutrality. Within months Turkey, Italy, and the other European nations were drawn into the war. Even Japan, on the other side of the globe, sided with Britain and seized several islands that were under German command.

The Great War soon became known for the unspeakable loss of human life it brought. In one battle in Belgrade in November 1914, Austria-Hungary lost 227,000 of 450,000 soldiers, and Serbia lost 170,000 of 400,000 combatants. In the next year, 1915, Britain lost more than 275,000 soldiers, Germany lost more than 600,000 soldiers, and France lost nearly 1.3 million soldiers. The continent of Europe was being exsanguinated of men.

On August 4, 1914, President Woodrow Wilson declared that the United States would remain neutral and uninvolved in the near madness of politics in Europe. Mindful of the country's many immigrants and their sympathies, two weeks later Wilson addressed Congress. "Every man who truly loves America will act and speak in the true spirit of neutrality," the president said. "The United States must be neutral in fact as well as in name during these days that are to try men's souls. We must be impartial in thought as well as in action." But there were limits to America's neutrality. As the war engulfed all of Europe, the United States was supplying the Allies with 400,000 rifles a year, 200,000 each from Remington and Winchester.

There was little feeling in the United States to join in the war. But in May 1915, Germany sank a British civilian passenger ship, the *Lusitania*, off the coast of Ireland. Nearly 1,200 people drowned, including 128 Americans. Despite the attack on American civilians, it seemed as if every American, with the notable exception of Teddy Roosevelt, wanted to keep the United States out of the war in Europe. During the Democratic national convention in 1916, William Jennings Bryan rose in support of the renomination of Wilson: "I agree with the American people," he said, "in thanking God we have a President who has kept—who will keep!—us out of war."

The governor of New York, Martin Glynn, rose to give the keynote speech at the convention, praising Wilson's ability to stay out of the war. "This policy may not satisfy . . . the fire-eater or the swashbuckler," he said. "But it satisfies the mothers of the land, at whose hearth and fireside no jingoistic war has placed an empty chair. It does satisfy the daughters of this land, from who bluster and brag have sent no husband, no sweetheart and no brother to the mouldering dissolution of the grave." The Democrats then gave the nomination to President Wilson and campaigned under the slogan, "He kept us out of war."

The Republicans nominated Supreme Court justice Charles Evans Hughes as their candidate. With hawks such as Roosevelt and Henry Cabot Lodge pressuring Hughes to come out in favor of sending American troops to Europe, the most Hughes could agree to was that America should enforce its neutrality. People began calling the Republican presidential candidate "Charles Evasive Hughes." Teddy Roosevelt chimed in that the only difference between the bearded Republican Hughes and the Democrat Wilson was "a shave."

As the campaign pushed into the autumn, its rhetoric intensified. The Democrats began insisting that electing Hughes would mean war, and they ran a full-page advertisement in many of the nation's newspapers that read, "You are working—not fighting! Alive and happy—not cannon fodder! Wilson and peace with Honor? Or Hughes with Roosevelt and War?"

In one of the closest elections in American history, Wilson defeated Hughes, aided by his insistence that he would keep the United States out of the Great War. Wilson took the oath of office for a second time in March 1917.

After the election results were known, the Germans reversed their earlier promise following the sinking of the *Lusitania* that they would not prey on American merchant ships. In January 1917, they announced that they would attack all shipping, neutral or not, civilian or not, that was bound for Britain. This was a violation of America's neutrality, but Wilson still did not ask Congress for a declaration of war. Then came one of the most curious events in American history.

In the first years of the war, the Germans had sought to stop the American arms shipments by sabotaging factories, but with little success. German saboteurs did manage to set fire to the Black Tom Shipyard in New Jersey in July 1916, but in general their efforts were comically incompetent. Franz von Rintelen, who had been sent to the United States to lead the sabotage efforts, quickly realized that this was not the way to stop the U.S. arms production. "I had studied the foreign policy situation of the United States and understood that the only country which the United States had to fear was Mexico," he later wrote. "Should Mexico attack the United States, the United States would need all the arms it can produce and would not be in a position to export arms to Europe."

At the time, relations between the United States and Mexico were at the lowest point since just before the Mexican-American War. Wilson had already sent American troops into Mexico twice in his first term, first in 1914 when he sent the U.S. Marines to Veracruz after American sailors were arrested.

Two years later, after Wilson recognized the new government of Venustiano Carranza in Mexico, the Mexican revolutionary Francisco "Pancho" Villa attacked Americans in Columbus, New Mexico, in retaliation. (What the Americans didn't know was that German instigators were encouraging Villa in the attack, even smuggling arms and other equipment to Villa in coffins and oil tankers.) Wilson then decided to invade Mexico, sending General John "Black Jack" Pershing on a futile attempt to capture Villa. These incursions into Mexico badly damaged the relationship between the two neighboring nations.

In January 1917, Arthur Zimmermann, the German foreign secretary, sent a telegram to the German minister in Mexico. The message said:

> *Telegram No. 1. Absolutely confidential. To be personally deciphered. We intend to begin unlimited U-boat warfare on February 1. Attempts will nonetheless be made to keep America neutral.*
>
> *In the event that we fail in this effort, we propose an alliance with Mexico on the following basis: Joint pursuit of the war, joint conclusion of peace. Substantial financial support and an agreement on our part for Mexico to reconquer its former territories in Texas, New Mexico, and Arizona. Settlement on details left to Your Right Honorable Excellency.*
>
> *Your Excellency shall present the above to the president [Carranza] in the strictest secrecy as soon as war with the United States has broken out, with the additional suggestion of offering Japan immediate entry to the alliance and simultaneously serving as mediators between us and Japan.*
>
> *Please inform president that unlimited use of our U-boats now offers possibility of forcing England to*

negotiate peace within few months. Confirm receipt. Zim-
mermann.

The Germans hoped that if the Americans were fighting a war along the Southern border, they would not have enough troops to send to Europe. What's more, Britain was getting a good amount of its petroleum from the oil fields in Mexico, and the German commanders knew that these oil fields would be one of the first targets of the Americans if there was a war.

"I don't think the Mexicans are in a position to take these areas [the American Southwest]," Zimmermann observed later, "but I wanted to hold them out in advance to the Mexicans as a goal, so that they would not be content to inflict damage upon the Americans on their own soil, but would immediately create incidents in the border states forcing the Union to send troops there and not here."

Originally the note was to be delivered to Mexico by the U-boat *Deutschland*, but because a U-boat required thirty days to cross the Atlantic, Zimmermann decided to send the coded message through U.S. diplomats. On January 16 the coded message was given to the American ambassador in Berlin, who transmitted it to the German embassy in Washington. It was forwarded on to the German diplomats in Mexico the next day.

Neither the Germans nor the Americans knew that the British had broken several of the German codes. The British secret service had a local spy in the telegraph office in Mexico City, and the British were able to decode Zimmermann's message after it arrived in Mexico. They handed it over to the U.S. State Department, and Secretary of State Robert Lansing personally delivered it to President Wilson. As Lansing read the short message to Wilson, three times the president exclaimed, "Good Lord!"

The message had fallen into Wilson's hands at an opportune moment: That week Congress was scheduled to vote on a measure that would allow merchant ships to be armed, therefore ratcheting the nation closer to war. Wilson supported the legislation, and he had Lansing call the senator heading the legislation and read him the telegram. Then the next day, Wilson told Lansing to release the telegram to the press. On the morning of the vote, papers across the nation headlined the news that Germany was trying to entice Mexico into a joint war against the United States. The Arming Bill passed the House by a vote of 403 to 13.

Supporters of Germany and opponents of the war quickly claimed that the message was a fake, reasoning that no nation would attempt such a clumsy strategy. Newspaper publisher William Randolph Hearst agreed, telling his editors that the telegram was "in all probability an absolute fake and forgery, prepared by a very unscrupulous Attorney General's very unscrupulous department." Despite the assurances of the Wilson administration that the note was real, many people, including many senators, refused to believe it. Then, on March 3, Zimmermann held a press conference and admitted that the telegram was authentic. He did so, he said, because "it will quickly occur to the American people what a dangerous position they could place themselves in by waging war against us."

As if sending the telegram wasn't enough, admitting to it in such a brazen and arrogant fashion angered many Americans. "As soon as I saw it," Senator Lodge wrote about Zimmermann's response, "I knew that it would arouse the country more than any other event."

Lodge was correct. The *Chicago Herald* reported, "Herr Zimmermann's Machiavellian invitation to Japan and Mexico, with whom we are at peace, to unite in the dismemberment of the United States despite its cold-blooded infamy, turns out to be of service to America. Congress still talks, but in a different key. The nation is aroused. The inexpressible cynicism and treachery of the German proposal has accomplished what the killing of American citizens failed to do."

Zimmermann's telegram had an especially strong effect in the Western states, which had been the most vocal in their opposition to entering the European war. There was no longer any question about whether the United States had a national security interest in seeing Germany defeated. The telegram was not the type of "overt act" that Wilson had talked about, but that was soon to come. Two weeks later German submarines sank three American merchant ships. On April 6, 1917, Congress passed a resolution of war against Germany, and the next day Wilson signed the resolution.

WHAT IF?

What if Woodrow Wilson had decided not to tell the public about the Zimmermann telegram?

"Good Lord!" President Woodrow Wilson said as Secretary of State Robert Lansing read aloud the intercepted message.

Lansing stood before the president waiting for instructions on how to proceed after this stunning change of events. Wilson sat behind his desk for several minutes, staring off into space and muttering, "Good Lord. Good Lord."

Finally Wilson looked at Lansing. "This means war. If the public were to know about this telegram and the threat to our national security, we would have no choice but to enter the war. Of course, this is little more than an insult, hardly more sophisticated than a schoolyard taunt. But nothing could inflame public passions more. We must not let any word of this leak out to the press."

Wilson stood and walked around his desk. "I want you to—yourself—send the strongest possible letter of objection. No, wait. That won't do—that would have to be sent to Zimmermann, and Kaiser William is the one I want. I'll do it. I will write a letter directly to Kaiser William. He has to know that if such threats continue, or if there is an overt act of violence against Americans, we *will* enter the war. I will plainly state that we will not enter the war on the side of the Allies, but as a third party, and that a peace settlement with the Allies will not extend to the United States."

Wilson's threat of a "war within a war" was soon conveyed to Kaiser William II. There were some in Wilson's administration who doubted that the approach would work, including Lansing, who thought that the president was being hopelessly idealistic in his constant pursuit of peace. But Wilson would not be put off. Wilson had a strong stubborn, self-righteous streak, and he insisted that he wasn't relying just on his own instinct that Kaiser William would react positively to the overture. "The time of the monarchies of Europe has ended," Wilson told his secretary of state. "Only fools resist Providence, and then to their own destruction. That we will prevail has already been ordained." Wilson wasn't just relying on his own interpretation of predestination, though. He had a suspicion that Kaiser William had been enticed into the war by his generals and admirals, and now that almost three-quarters of a million German soldiers had lost their lives with no end to the war in sight, the Kaiser would consider a settlement of an honorable peace.

When the German emperor dismissed Arthur Zimmermann from his post as foreign secretary, using diplomatic smoke signals to let Wilson know of his acceptance of the warning, the president knew that Kaiser was willing to listen.

Wilson began secret negotiations with the Kaiser to end the war before the United States was forced to become involved. Wilson found Kaiser William vacillating and somewhat immature for a man of his age and stature, but Wilson, the son of a Presbyterian minister, was convinced that he could save the Kaiser's political soul. In March 1917, the three-hundred-year-old reign of the Romanov family in Russia had resulted in a violent and bloody coup in which the entire Romanov family had been murdered. The Austro-Hungarian Hapsburg Empire was certain to fall. The only major monarchy that still existed on the continent was Germany's Wilhelm Imperial Empire.

With appropriate but atypical deference, Wilson pointed out that the longer the war dragged on, the greater the risk that Kaiser William's reign could end as well.

Having put Germany on notice, Wilson knew that the next provocation, no matter how inconsequential, would cause him to commit troops to Europe. Even if an undisciplined German submarine captain mistakenly fired a torpedo at an American ship, Wilson would have no choice but to declare war. He did not want a lone German naval officer deciding the fate of the United States, and so he decided to change his policy of neutrality. Wilson took the unusual step of going to Capitol Hill to address the

Senate, to explain his change in policy. The situation in Europe must end, he told the senators, in "peace without victory."

"Victory would mean peace forced upon the loser, a victor's terms imposed upon the vanquished," he declared. "It would be accepted in humiliation, under duress, at an intolerable sacrifice, and would leave a sting, a resentment, a bitter memory upon which terms of peace would rest, not permanently, but only as upon quicksand."

Wilson announced that he would embark on a plan of "Real Neutrality" (much like that which had been proposed by his first secretary of state, William Jennings Bryan), in which all participants in the war are equals. The United States would join other neutral nations, Denmark, Sweden, the Netherlands, and Norway, in eliminating the sale of munitions and war materiel to any of the belligerents. Wilson's unstated hope was that eliminating the transport of war materiel to the Allies would likewise eliminate Germany's presumed need to attack American ships.

Wilson went further. In 1916 he had opposed the Gore-McLemore Bill, which would have prohibited passenger travel into the war zone, but in this address to the Senate, Wilson called for just such a prohibition. He was concerned that American merchant ships might begin carrying passengers to serve as human shields against German attacks. Such a scheme would almost certainly end in a tragedy, one that would force the United States into the war. It was better to temporarily limit some of the travel rights of Americans, Wilson decided, than to create a situation where war was the only option.

In the end, mentioning the many immigrant groups

in the United States, Wilson asserted that his neutrality plan was "the only sort of peace the peoples of America could join in guaranteeing."

Late that evening, Wilson met with an old friend, Frank Irving Cobb, who was an editor from the *New York World*, in the White House. "Do you know what American intervention would mean?" the president asked his guest. "Germany would be beaten and so badly that there would be a dictated peace.

"It means," Wilson continued, "an attempt to reconstruct a peacetime society with martial standards. And, at the end of the war, who would be willing or with sufficient power to enforce the terms? A dictated peace would be no peace. There won't be any *peace terms* left to negotiate with. There will only be *war terms*."

Conscious of the possibility of America entering the war, and weary of the heavy losses, Kaiser William and President Wilson worked out a peace settlement that Wilson then presented to the governments of Britain and France.

In the famous Twelve Points Peace Settlement, Germany agreed to withdraw from France and Belgium. In exchange, Germany was allowed to annex Lithuania, Luxembourg, and the French territories of Briey and Longwy. Germany also acquired the territory of the Belgian Congo. French Morocco was given its independence, which had been a German demand in the years before the war. To quell the militaristic ambitions of Germany's ruling class, Wilson persuaded Kaiser William to accept political changes in Germany as well. The German chancellor was no longer chosen by the Kaiser but instead by Germany's lower parliament, the Reichstag. This change, which had been advocated by liberal members of Germany for decades, had the effect of making Germany more completely a democratic nation and placing the German army and navy under civilian command.

Woodrow Wilson won the 1919 Nobel Peace Prize for bringing about an end to the Great War, but more important to the people of the United States, he had kept them out of the European conflict. Wilson, who became known as America's greatest peacekeeper, created an environment of peace and prosperity in Europe that lasted for decades to come.

Of course, Wilson did not keep America out of the Great War, and his decision cost many lives. On a single day in May 1921, the bodies of 6,000 American soldiers arrived home from Europe—in all, 170,000 American soldiers lost their lives fighting in World War I. (After the soldiers returned home, another 400,000 people in the United States died from a strain of influenza that was imported from the battle trenches.) In Europe, an estimated 20 million people were killed by the war.

In November 1917, Wilson drew up a peace plan, called the Fourteen Points, without consulting Britain or France, and Germany agreed to this chance to end the war. When the British and French began changing portions of the settlement, Wilson agreed to their demands, without renegotiating with Germany. The Germans and the new Soviet government were not allowed to participate in the negotiations at Versailles, and the Germans felt betrayed by the altered language of the treaty and by the fact that they had not been allowed to participate in the negotiations. But with the Allies holding strategic parts of Germany, the leaders of the country believed that they had no alternative but to resentfully agree to sign.

Some Americans were dismayed that the Germans had been shut out from determining their own future. William Bullitt, who would later become the U.S. ambassador to France and Russia, wrote to President Wilson, "Our government has consented now to deliver the suffering peoples of the world to new oppressions, subjections and dismemberments—a new century of war."

Chapter 19

Retaining Ruth

❧

The Boston Red Sox traded the most famous player in base-ball. But what if the Sox had kept the Bambino in Boston?

MILLER Huggins, manager of the New York Yankees, had been given a difficult task. He had been sent to Los Ange-les to find Babe Ruth, who was on the West Coast enjoying his newfound celebrity. For three days Huggins searched fruit-lessly for the baseball hero.

Ruth, a star for the Boston Red Sox, had just set the major league record for home runs the season before with an astound-ing twenty-nine four-baggers (surpassing the 1884 mark of twenty-seven set by Ed Williamson of the National League's Chicago team). He had gone to Los Angeles after the season, making short films that exploited his celebrity and wowing the locals with 340-yard drives on the golf course. At the Griffith Park golf course, Huggins finally caught up with Ruth as he came off the 18th green. As soon as Ruth saw him, he had an inkling of why Huggins was there, and asked, "Have I been traded?"

Ruth understood immediately what Boston baseball fans

still can't fathom. The best player in the sport had been sold to their arch rivals, the New York Yankees. It was a trade that not only changed the fortunes of those two teams but also changed the sport of baseball—and America, too.

❦

Even though the baseball lords had introduced a farther-traveling corked baseball in 1910, the game was a slower, more plodding affair than has been seen in the modern era. Baseball was played in what is now known as "small ball," which means moving the men around the bases one base at a time. A player might reach first on a single, be moved to second on a bunt, steal third, and then score on a sacrifice fly to win the game one to zero. It was a time when taking a mighty swing and whiffing was a disgrace, as depicted by the popular baseball poem "Casey at the Bat." A batter was expected to choke down on the bat, control his swing, and make contact with the ball at each trip to the plate.

Ruth entered the major leagues in 1914 as a pitcher for his hometown Baltimore Orioles, but the team traded the rookie pitcher to the Boston Red Sox that first summer. Ruth soon became dominant on the mound, leading the league with the lowest earned-run average in baseball (1.75) two years later. As a pitcher for the Red Sox in the 1916 and 1918 World Series, Ruth pitched a record $29^2/_3$ scoreless innings. As a pitcher Ruth didn't play every day, and in keeping with baseball strategy, the manager sometimes sent a pinch hitter to the plate when Ruth's turn to bat came up in the order.

In 1918 the Red Sox hired a new manager, who decided to place Ruth in the outfield on the days that he wasn't pitching so that he could take advantage of Ruth's hitting abilities. Ruth responded by tying for the league-best total with eleven home runs. Ruth swung at the ball the way he lived life: with enthusiastic abandon. The next year he broke the major league record for home runs in a season with twenty-nine, a feat that the *New York Times* said would "stand for many years to come." The Red Sox had a star player, but more important for the team's owner, Harry Frazee, the team now had an asset that he could put on the auction block.

Frazee was in the entertainment business, and besides owning a major league baseball team, he had backed several unsuccessful Broadway plays. *Ready Money* failed to make him rich in 1912, *A Pair of Sixes* wasn't a winning hand in 1914, and *A Full House* produced only empty seats in 1915. Frazee was badly in need of money, so he sold Babe Ruth to the Red Sox's rivals, the New York Yankees, for $100,000 plus a $350,000 loan (with Fenway Park as the collateral).

When Frazee sold Ruth to the Yankees, baseball fans in Boston were beside themselves. An editorial cartoon appeared in the newspaper showing "For Sale" signs on Faneuil Hall and the Boston Public Library. Frazee imperiously claimed that Ruth was the reason that the Red Sox had finished in sixth place in 1919 and asserted that allowing Ruth to stay in Boston would have been an "injustice," because the Sox were in danger of becoming a one-person team. "Ruth had become simply impossible, and the Boston club could no longer put up with his eccentricities," Frazee said. "I think the Yankees are taking a gamble. While Ruth is undoubtedly the greatest hitter the game has ever seen, he is likewise one of the most selfish and inconsiderate men ever to put on a baseball uniform." The baseball sports writers in Boston were caught up in Frazee's spin and began writing that Ruth was washed up.

Frazee told the press that he was going to use the money to build a team that could contend for the pennant, but this was a lie. In fact, Frazee needed the money to pay off his debts from the failed shows. When he put up posters for one of his new shows the next season, a play called *My Lady Friends*, a bitter fan pointed at the poster and said, "Those are the only friends that son of a bitch has."

The Ruth deal was twice as much as had ever been paid for a baseball player before, and it was controversial in New York, too. "Columbia had to pay no such sum to the University of Chicago for the release of an eminent professor," sniffed the *New York Times*, but the paper admitted that the deal probably made business sense because the Yankees were likely to pull in enough fans to make the transaction profitable.

Neither the *Times* nor the Yankees owner, Colonel Jacob Huston (who had a strong German accent and always referred to his star as "Babe Root"), had any fantasy that the deal would work out as well as it did. The next season Ruth obliterated his own record of twenty-nine home runs for the season by hitting an amazing fifty-four. At the time the career record for home runs stood at just 119, and Ruth passed it in the middle of the following season. With the Yankees, Ruth became the first player to hit thirty home runs in a season. He then became the first player to hit forty home runs in a season. Then the first to hit fifty. Then, in 1927, the first to hit sixty.

Ruth became almost as famous for off-the-field exploits as he was for his heroics between the chalk lines. One reporter captured the image of Ruth as "a large man in a camel's hair coat and camel's hair cap, standing in front of a hotel, his broad nostrils sniffing at the promise of the night." When a reporter asked Ruth's road roommate what it was like to room with the Bambino, he famously replied, "I don't room with him. I room with his suitcase."

The Ruth-led Yankees saw their attendance jump by 20 percent in the first year he was on the team, and soon sellouts were commonplace. The Yankees became the first team to draw more than a million fans in one season, a mark that they achieved in the middle of Ruth's first season with the team. In 1923, the Yankees opened a new stadium to hold all of the fans. It truly was the house that Ruth built, as it would later be pegged, because going to see Ruth play was the most popular tourist activity in the nation.

Meanwhile, the Red Sox were suffering, in part because of their loss of Ruth. The team finished in last place nine times in eleven seasons, and Frazee was eventually forced to sell the team, although he finally did get a Broadway hit in 1925 with *No No Nanette* and its hit song, "Two for Tea." The Red Sox, many fans believed, were jinxed for selling the best player in the history of the sport, and as the Sox lost the World Series in the seventh and final game four times over the next sixty years, fans were still cursing the trade decades later.

Ruth and the Yankees went on to play in eight World Series, and the 1927 Yankees' team is considered perhaps the best baseball team of all time. In that year, Ruth set his famous record of sixty home runs and Lou Gehrig added another forty-seven, together accounting for almost one-fourth of all of the home runs hit in the major leagues that year. When Ruth finally quit the game in 1934, he had hit 714 home runs. At the time, only two men had ever hit more than 300 home runs in a career. Ruth's record stood until Henry Aaron passed it forty years later, and his record of sixty home runs in 154 games stood until Mark McGwire surpassed it in 1998.

WHAT IF?
What if Harry Frazee had decided to keep Ruth in Boston?

In the fall of 1920, baseball was hit with a lightning strike when eight players of the Chicago White Sox confessed that the previous year they had thrown the World Series, intentionally losing to the underdog Cincinnati Reds. The Black Sox, as the eight became known, were banned from baseball by Commissioner Kenesaw Mountain Landis. Frazee and the other owners were worried about what the scandal might mean to baseball.

Attendance at the ballparks had been slipping during the past few seasons, and many of the baseball owners blamed the automobile. People had gone crazy over Henry Ford's Model T, and instead of coming to the ballpark, they were spending their weekends and evenings driving to see friends and driving to visit relatives and going for drives just for the fun of it. Those black Tin Lizzies were scurrying everywhere, Frazee thought, like mindless ants scurrying around an anthill. Those who weren't driving and driving were working evenings and weekends so that they could afford to buy one of the things.

As if that wasn't enough to drive a man to distraction, now the hero of his baseball team, that baby Ruth, was demanding that Frazee double his salary. He had signed a contract and he had two years still to go. And immediately after he did so, he had gone to Los Angeles to make movies. Movies! Frazee thought. The very thing that was ruining attendance at his Broadway plays. The man was simply insufferable!

All he needed to make his Broadway shows work, Frazee thought, was a star, a big name who would make people come back to the theater. A star who would grab headlines across the nation, a star who would make the people want to get in those cars and drive to the theater. One star, and his worries would be over.

Frazee was drowning in debt when Colonel Huston offered to buy Ruth's contract. It was enough money to clear his debt and get his theatrical productions back on sound footing. But as Frazee thought about the Yankees' offer and his need for the dollars a star would bring in, the obvious struck him. He already had a star, one that people were willing to go to the movies to watch, one that everyone wanted to see. His star wasn't on the stage but on the ballfield. That was why Colonel Huston was so eager to pay all of that money for Ruth, Frazee realized. If Huston is willing to pay nearly half a million for Ruth, Frazee thought, he had to be worth several times more than that. It was a gamble, but what show wasn't? With Ruth, Frazee could put on the biggest show in America—and in one of the biggest venues in the country.

Frazee's hunch proved right. In the 1921 season, fans turned out by the thousands to see Ruth play, and the Boston Red Sox turnstiles spun so quickly that Frazee's club set an attendance record, becoming the first club to attract more than one million paying customers. Ruth didn't disappoint the fans, either, and he again won twenty games as a pitcher and also became the first player in major league baseball to hit thirty home runs for the season.

As the season entered the fall playoffs, the Ruth-led Red Sox once again went back to the World Series against the New York Giants. Before the series began, Frazee was approached by a group of men from Newark, New Jersey, who had a plan to broadcast the game to the public. A device had been invented fifteen years earlier, a type of wireless telegraph that could carry voices, and it had been used by ships and by technically oriented hobbyists.

These men wanted to broadcast the Series from their station WJZ, which was one of just eight radio stations in the country, to anyone in New York or New Jersey who cared to listen in. Frazee was flabbergasted.

These men actually wanted to deliver the news of the game to people for free! If he allowed this, Frazee thought, why would anyone want to come out to the ballpark? Baseball was already struggling to overcome the Black Sox scandal and the diversion of the automobile, and now these men hoped to just give away what was left! And if this radio device became popular—one of the promoters, a man named David Sarnoff, was talking about selling a radio receiver for each home—what would this mean to Broadway? What if they asked to transmit Broadway plays next? This was a bad idea, Frazee told his fellow team owners, and he wasn't going to allow it. The Baseball Writers Association agreed, sending a telegram to Commissioner Landis saying that allowing radio to broadcast the games not only would kill attendance at the ballparks but might ruin some newspapers, too. The team owners soon decided that radio had no place in baseball.

Babe Ruth played another six years in Boston, leading them to three championships. He was one of the greatest pitchers in the game, but he was best known for his exploits at the plate on the days when he wasn't pitching. After he retired, Ruth became one of the initial inductees in the Baseball Hall of Fame.

The game of baseball, though, struggled until after World War II. Some GIs who had been stationed in Great Britain had seen a new device that transmitted pictures and sound through the air, a device that the British Broadcasting Corporation had used to broadcast the coronation of King George VI in November 1937. When the soldiers returned to the United States and began buying televisions, the baseball owners decided to allow the television station owners to broadcast their games in exchange for a fee. Although some owners had been as resistant to allowing television into baseball parks as they had been to radio twenty-five years earlier, they were delighted to discover that television actually increased the number of baseball fans, and baseball became more popular than ever.

The sale of Ruth to the New York Yankees had a major effect on baseball. His hitting prowess changed the way the game was played, and his popularity did more than just bring fans back to baseball.

Computer companies often search for what they call a "killer app," a software application so attractive to consumers that they will buy new computers just to be able to run the new software. For the owners of radio stations in the early 1920s, Babe Ruth was the killer app. In 1921, the World Series between the Giants and Ruth's Yankees was broadcast by WJZ in New Jersey. (A Newark newspaper reporter called in the action over a telephone, and the play-by-play was repeated over the air by an announcer.) People gathered in homes that had radios, and in streets outside businesses that had set up speakers, to listen to the Series and to cheer for Ruth. The next year, $60 million worth of radios were sold; by 1929 that amount had risen to nearly $1 billion, and part of the popularity of radio was due to Babe Ruth and his ability to hit home runs like no human before.

From the 1920s on, listening to a baseball game broadcast on the radio became a favorite American pastime, as much a part of our culture as enjoying a hot dog at the ballpark. Without Ruth, and Henry Frazee's unfortunate decision to sell him, listening to ballgames on the radio might have been a pleasure that never would have been known.

Without Ruth, the Yankees probably would still have won championships in the 1920s, but they never would have acquired the aura of one of the greatest sports teams ever. And without Ruth, the Boston Red Sox suffered through what many have described as a curse caused by the trade of the Bambino. Four times—in '46, '67, '75, and '86—the Boston Red Sox made it to the final, championship game of the World Series, only to come up short. The last time the Red Sox were able to capture the crown was in 1918, the last year Ruth played on the team.

Chapter 20

The Phony War Becomes Real

Hitler and Stalin joined together to divide Eastern Europe for themselves. What would have happened if the United States had joined Britain in resisting this assault on sovereign states?

THE early part of the twentieth century saw the great idea of the eighteenth century, democracy, come under worldwide assault. Aggressive, militaristic governments were found in Mussolini's Italy, Franco's Spain, Stalin's Soviet Union, Hirohito's Japan, and Hitler's Germany, and it was this type of government, not democracy, that seemed to be winning.

In August 1939, the world received shocking news: Two oppressive, totalitarian regimes, Nazi Germany and the Soviet Union, had joined hands in what was termed a "mutual nonaggression" pact. Europe was about to become a very dangerous place.

The Great War had done more than ruin Germany politically and militarily—it had also destroyed Germany's economy. At the end of World War I the German mark had been worth one-fourth of the U.S. dollar. By 1923, just five years later, the value of the mark had fallen a billion times after the worst economic inflation ever seen in modern times. Some people in Germany were starving.

During the economic collapse in Germany in the 1920s, many Germans were desperate for a way out of the crisis. Some turned toward Communism, some favored Socialism, and some the fascist and ill-named *Nationalsozialistische Deutsche Arbeiterpartei*, the National Socialist German Workers Party, or Nazi party. The leader of this group was an Austrian-born former corporal in the German army, Adolf Hitler.

The Nazi Party emphasized the superiority of Germany and the German race and predicted that although the nation was humiliated, its economy in ruins, and its people hungry, Germany would one day rule the world, an argument that had been made in Hitler's book, *Mein Kampf* ("My Struggle"). The groups who opposed this vision, this destiny of Germany, the Nazis believed, were the Jews, the Socialists, and the Communists. When the Nazi party assumed power in Germany in 1933, many Jews and Communists were arrested and tortured. Hitler had written in *Mein Kampf*, "We must never forget that the regents of present-day Russia are common blood-stained criminals; that here is the scum of humanity."

Adolf Hitler and many Germans in general believed that their nation's difficulties began with the unjust Treaty of Versailles. Almost immediately after becoming chancellor in 1933, Hitler began violating it. The treaty restricted Germany to an army of just one hundred thousand men, but in 1935 Hitler announced that Germany would no longer abide by that limit. In 1936 he sent German troops into the restricted Rhineland area and began to build fortifications in Germany; both actions violated the treaty. Britain and France said that they would go to war again to enforce the Treaty of Versailles if the United States would agree to participate, but President Franklin Roosevelt refused.

Despite the saber-rattling, in 1938 France remained Germany's top economic trading partner. Germany's relations, economic and otherwise, with the Soviet Union were much cooler than with the French. "The rumor that our Fuhrer will someday join hands with the Jewish Bolsheviks of Soviet Russia must have originated in the filthy political sewers of Paris," said an editorial in the German periodical *Essener Nationalzeitung* in December 1938. "As long as we remain National Socialists, true to our ideals, there can never be any commerce between Nordic Germany and Asiatic Russia."

Hitler demanded that Czechoslovakia turn over the German-speaking territory of Sudetenland to Germany. Eduard Benes, president of Czechoslovakia, asked France and Britain to help defend his country against the predatory Germans in March 1939. But French premier Édouard Daladier and British prime minister Neville Chamberlain, whose countries were committed to defend Czechoslovakia, refused to live up to their word. In a 1939 meeting in Munich, they tried to appease Hitler by giving the territory over to Germany. "You have gained shame and you will get war," said an agitated Winston Churchill.

The British and French governments asked the Soviet Union to come to the aid of the Czechs, but Stalin wasn't interested in attacking Germany, especially with two allies who were weak-kneed. When Stalin refused to join the British and French, rumors began that the Nazis and Soviets would soon join in an alliance, even though the leaders of both counties dismissed the idea. "The whispered lies to the effect that the Soviet Union will enter into a treaty of understanding with Nazi Germany is nothing but poison spread by the enemies of peace and democracy, the appeasement mongers, the Munichmen of fascism," wrote the *Soviet Daily Worker* in May 1939.

France and Britain had sent signals that they were interested in joining the Soviet Union in fighting Nazi Germany, but Stalin said that he preferred the "practical arithmetic" of signed agreements to the "algebra" of diplomatic signals. Neither France nor Britain was willing to sign an agreement with Communist Russia.

In August 1939, the Soviet commissar of foreign affairs, Vyacheslav Molotov, and German foreign minister Joachim von Ribbentrop signed a nonagression treaty. Stalin was not yet ready to go to war with Germany, and he decided that the only way to avoid a war was to sign a treaty instead. "For six years German fascists and the Communists cursed each other," Stalin said about the treaty. "Now an unexpected turn took place; that happens in the course of history." The treaty may have been termed a nonaggression treaty, but the nonaggression applied to the Nazi-Soviet relationship only. The countries in Europe had much to fear. In fact, the treaty said that the two countries would divide Poland between them. Furthermore, the Soviet Union could have Finland, Romania, and the Baltic nations, and Germany could take the rest of the Europe. The treaty also bartered Soviet raw materials for German manufactured goods.

Hitler wanted to punish the Western European countries for what they had done to Germany in the years following World War I, but he knew that he could not do so without attacking Poland first. Poland was considered a major military power in Europe, famous for its mounted cavalry, and although Hitler planned to attack France, he did not want to fight a two-front war. He decided to invade Poland before he had to fight France and Britain. Many observers thought Germany could not defeat Poland. Military writer Maj. George

Fielding Eliot said in the *Boston Evening Transcript* in May 1939, "The chances of Germany making a quick job of overwhelming Poland are not good."

But the Germans had developed a new strategy of war, the blitzkrieg, which made full use of modern military machinery. Instead of sending in broad lines of infantry backed by tanks and artillery, Hitler sent in narrow lines of tanks, like arrows shooting into an animal, protected along their flanks by Stuka airplanes. The tactic had its detractors: A June 1939 article in *Time* magazine said, "The modern German theory of victory by Blitzkrieg (lightning war) is untried and, in the opinion of many experts, unsound."

With the nonaggression treaty in hand, Hitler invaded Poland. After the Nazis faked an attack on a German radio station near the Polish border, the Germans rushed in with 250,000 soldiers protected by 1,600 Luftwaffe airplanes on September 1, 1939. The attack on Poland contained a warning to the United States, too. One of the first German bombs that dropped on Warsaw hit its target, a villa that the American ambassador had rented for the staff of the American embassy.

Two days after Hitler invaded Poland, Britain and France declared war on Germany. That same day, German foreign minister von Ribbentrop told the Soviet Union that the eastern half of Poland was theirs for the taking, and on September 17, 1939, the Soviets joined the Nazis in invading Poland. It was the first time that the Soviet Union had waged war to gain territory.

The Polish people waited for the Western powers to save them. Throughout the day the Polish radios played Chopin's "Military Polonaise" interspersed with appeals for immediate assistance from Britain and France. Although war had been declared, no real help was coming, and the hell of totalitarian occupation was just beginning. In eastern Poland, Hitler began rounding up Jews, Gypsies, and Socialists and sending them to concentration camps to await their executions. Meanwhile, in western Poland, Stalin ordered Polish military officers arrested. They were, and in the spring of 1940 in Katyn Forest, all 15,000 of them were murdered as Stalin had commanded.

Poland had signed a treaty with France that specified that the French would send troops within days if Germany attacked, but despite the declaration of war, no French troops fired any weapons. Using rail lines within sight of French artillery, German trains carried supplies to troops in Poland, and not a shot was fired. Likewise, the British war effort in the autumn of 1939 was nonexistent. British bombers did fly over Germany after the invasion of Poland, but instead of releasing bombs, the planes dropped propaganda leaflets denouncing Hitler. An American senator, Arthur Vandenberg, gave the events of the fall a label that would stick, calling the conflict the "phony war." "This so called war is nothing but about twenty-five people and propaganda," he said.

Then, in November 1939, the Soviet Union invaded Finland, expecting to take control of that country as easily as it had Poland. But the Soviet army had a problem: Stalin, in his paranoid and horrific measures to maintain power, had executed nearly every capable top officer within his own military. The Soviet army was large but poorly trained and equipped, and it lacked experienced leaders. Finland did not fall easily, but instead skillfully used its ski patrols to destroy Soviet supply lines and to surround Soviet troops. "For four miles the road and forest were strewn with the bodies of men and horses," one witness wrote of one Finn victory. "The corpses [of the Soviet soldiers] were frozen as hard as petrified wood." After three months of fighting, the Soviets were able to use bombers to overwhelm the Finns. At the end of the ordeal, Stalin said, "The friendship of the peoples of Germany and the Soviet

Union, sealed in blood, has every reason to be lasting and firm."

With the Allies' phony war doing nothing to dissuade Hitler from further aggression, in April 1940 he invaded Norway and Denmark. The British rushed to Norway with aid, but the force was too late and too little, and Norway and Denmark quickly fell.

As much as the politicians in Britain and France wanted the status quo to hold, the phony war could not last. On May 10, 1940, Winston Churchill, a politician who had been considered an alarmist for his braying about the threat of Hitler, became prime minister of Britain. On the same day, Hitler invaded Belgium, the Netherlands, and Luxembourg, all of which quickly collapsed. With the Low Countries under his control, Hitler simply sent his tanks and motorized infantry around France's fortified Maginot Line by driving through Belgium. The French had placed most of their two-million-man army in the catacombs of the Maginot Line facing Germany. But Hitler invaded from Belgium, not from Germany, and he faced little resistance. Churchill was forced to abandon his support of France by withdrawing his troops, and Britain barely evacuated its army from the beach at Dunkirk before the Nazis could destroy them.

Just as Poland had pleaded vainly for a rescue from Britain and France, now these two countries pleaded with President Roosevelt (FDR) to come to their aid.

The United States was not well prepared to help Europe. During the 1930s its economy was in a state of collapse, and although no treaty required it, as was the case with Germany, the U.S. had unilaterally reduced its standing army to under 100,000 men (it had been 1,200,000 just fifteen years earlier).

Roosevelt decided not to send troops to Europe, and he and Hitler engaged in mutual avoidance. FDR abandoned Woodrow Wilson's policy that neutrality meant being able to

sail the seas unmolested, and he restricted travel. Hitler
ordered German U-boats not to attack American ships, and
the German press stopped criticizing America and the Roosevelt administration. The German embassy to the U.S. also
paid for people with antiwar beliefs to attend both the Democratic and Republican presidential conventions, and they quietly
helped fund editorial writings by isolationists such as Charles
Lindbergh.

Adolf Hitler thought that Germany could avoid a conflict
with the United States for decades. In 1940, he told Soviet foreign minister Molotov, "The United States will not be a threat to
us for decades—not in 1945, but at the earliest in 1970 or 1980."

WHAT IF?

What if Franklin Roosevelt and the U.S. Congress had pre-
pared for a possible war and then interceded when Germany
invaded France?

On November 3, 1936, Franklin Roosevelt was in
New York to vote for president. He had given a rousing
speech the evening before in Madison Square Garden, a
speech full of anger at the Republicans and their absurd
accusations about the constitutionality of his policies that
they had thrown out near the end of the presidential campaign. After casting his vote, there was nothing for him to
do but wait for the election returns.

He won, of course. Alf Landon, the Republican candidate, had no chance of unseating him. The nation's
economy was improving, and as long as that was the case,
the incumbent president rarely had to worry about reelection. But as he sat alone in his first moments of peace in

months, waiting for the first election returns, Roosevelt allowed himself to think about something that he had tried to put out of his mind for months. What was to be done about the German chancellor Adolf Hitler?

As an old navy man, Franklin Roosevelt understood what a revitalized German military could mean to the world's merchant fleets. Europe in 1936 was a very dangerous place. This man, Adolf Hitler, had flagrantly ignored the Versailles Treaty after gaining office. None of the nations that had signed the agreement—Britain, France, the Netherlands, or the United States—had any desire to enforce it and possibly return to the trenches of the previous war. There was little in the world more horrible than a repeat of the Great War, Roosevelt thought. But there was an increasing danger that if the Allies weren't prepared to stop Hitler's violations of the treaty, another world war would result.

How would it begin? Most probably as the Great War had begun, with German submarines creating havoc in the Atlantic shipping lanes. That would interrupt American exports, and the U.S. economy was in no condition to absorb this kind of assault. FDR believed in Wilson's philosophy that America's neutrality entailed its ships traveling unmolested on the high seas, but such beliefs were empty without the might to enforce them. Roosevelt decided that in the next few years he would need a powerful navy to enforce the United States' neutrality in any approaching European conflict. "You don't wait until a snake bites you before you kill it," he had told his cabinet a week later.

Roosevelt decided that he could solve two problems with one solution. As Ronald Reagan would do fifty years later, Roosevelt spent his way out of an economic squeeze with a military buildup, one that would begin with a massive program of shipbuilding.

Roosevelt realized that such a program to build ships, as well as planes, tanks, and trucks, would put millions of Americans back to work. Many in his own party would oppose the plan, but Roosevelt knew how to present the news. In a fireside radio chat, he announced the major military buildup. "Our sons, daughters, and sweethearts should never again have to fight for a foreign power," he said. "But if men of evil intent decide that the United States isn't ready, they will harm us on the high seas and sink our world reputation just as they did before the Great War. I promise you tonight that this will not happen again. Never again, never again, never again." In the speech, FDR announced that he was also going to call for a half-million soldiers and sailors to be added to the nation's armed forces, bringing the total to six hundred thousand. "While still short of the great armies of Europe, this will allow us to respond to any provocation indefinitely."

Critics attacked Roosevelt's plan immediately, but as the stock market responded favorably the next day, and as articles began appearing in the newspapers about lines of job applicants at shipyards, steel mills, factories, and even at the local diners that served these industrial giants, the critics' voices became silent.

By 1939 the American military was the strongest that it had been in twenty years. Neville Chamberlain said that he didn't want the United States involved in European affairs, and Roosevelt was relieved, because he had had no intention of offering the British any assistance should they go to war with Germany over Czechoslovakia. But world affairs would soon change both nations' plans.

In June 1940, Nazi Germany invaded France, and on

June 14, 1940, French premier Paul Reynaud asked Roosevelt to send troops to Europe in France's hour of need.
Winston Churchill also asked Roosevelt to send troops to
stop the Nazi juggernaut.

Roosevelt decided that he had to act to stop Hitler,
but there was work to be done before he would commit
fighting men. First was the situation in the Pacific. Japan
had been aggressively inserting itself into the affairs of
every country in Asia and, of course, had invaded China.
Roosevelt had a visceral dislike of the militaristic government of Japan. However, he could not conduct a war
against Japan and still afford to send troops to Europe.
Roosevelt sent a cable to Churchill:

> *Secret from the President to the Former Naval Person*
> *On June 16 I would like to present to the Japanese*
> *Ambassador a proposal for a* modus vivendi *[a temporary*
> *working agreement] that might give us an opportunity to*
> *develop a comprehensive program of peace covering the*
> *Pacific area and to relieve the domestic political situation in*
> *Japan. I would suggest that we work together with France,*
> *Belgium and Australia to supply Japan with a required*
> *quantity of petroleum products and other necessities pro-*
> *vided that no nation makes an armed advancement in*
> *southern Indochina. This seems to me a fair proposition for*
> *the Japanese. Roosevelt*

What went unstated was that the British would have
to share their petroleum reserves in Indochina with the
Japanese, but Churchill recognized the geopolitical necessity of the agreement. The second request that Roosevelt
put to Churchill, however, turned out to be a bit more
problematic than Roosevelt had anticipated.

Roosevelt knew that the British had one weapon unequaled in the modern world, and that was Winston Churchill's mouth. FDR still faced the difficulty of overcoming America's entrenched isolationism, and he asked Churchill to come present the case for American intervention to a joint session of Congress. (Like much of what Roosevelt did, it was a wily and sophisticated political move. If he was to make the case himself and it was rejected, the stain would be on him. If Churchill failed, then Roosevelt was just a bystander. If Churchill was successful, though, Roosevelt would receive credit for orchestrating the United States' involvement in the war.) Churchill initially balked. After the near-disaster of Dunkirk, Churchill felt that he needed to stay in England. But after a week of urgent cables between the two men, Churchill agreed to make the request for American troops in person.

Churchill's oratory did not disappoint Roosevelt. Speaking before an assembly of the House, Senate, and Supreme Court justices, and before a worldwide audience via radio, Churchill presented the reasons why the United States had to aid Europe. "Britain is fighting by itself alone, but we are not fighting for ourselves alone," Churchill said. "This is no war of chieftains or of princes, or dynasties or national ambition; it is a war of peoples and causes. There are vast numbers, not only in this land but in every land, who will render faithful service in this war. Their names may never be known, but their deeds will be recorded for eternity."

After the tremendous response to Churchill's speech, two days later Roosevelt delivered to Congress his own message asking for a declaration of war against Germany and the Soviet Union: "If Britain and France fall, the

United States will be the last, lone democracy in an unfriendly and dangerous world. If we do not stand up to this unholy alliance, the light of liberty will be dimmed throughout the world."

When Britain, France, and the United States declared war against Germany and the Soviet Union on June 21, 1940, Soviet premier Joseph Stalin realized that he had been caught on the wrong side of the line with the wrong ally. That suspicion was confirmed when Japan decided to take advantage of the situation in Europe and attacked the Soviet Union, with hopes of capturing the rich Siberian oil fields.

Relations between the Japanese and the Russians had been poor for decades. The Japanese were angry over Soviet aid to communist rebels in China, and the countries had a dispute over Sakhalin Island, with the Soviets insisting that Japan give up the southern half of the island and Japan insisting that the Soviets share the oil and coal in the northern half. In the summer of 1938 the Soviets fought off an attack by Japanese forces on the Manchurian border, and in May 1939, there had been fighting between Russian and Japanese troops in which the Japanese were badly beaten.

Many of Japan's military officers considered the Soviet Union the main threat in Asia. The Japanese launched a full attack on the Soviet Union in September 1939, and within a week had captured all of the Sakhalin Island and Vladivostok. By September 1940, Japanese forces had moved north through Manchuria to capture the Soviet city of Irkutsk and controlled most Russian territory west of the Lena River.

After a year of fighting alongside Germany in Europe and against Japan on the steppes of Siberia, in October

1940 Soviet Union announced that it was withdrawing from the alliance with Germany and removing its troops from Finland. The Soviets were successful in pushing the Japanese army out of Russia (with the exception of Sakhalin Island) by May 1941, but the militaristic government of Japan continued to cause problems in Asia for the next thirty years. Whether through geopolitical pragmatism or subtle racism, Americans never considered Japan's aggression in Asia as repugnant as the Nazis' aggression in Europe in the 1940s.

In August of 1941, the Allies achieved a negotiated surrender with Germany. All of France had been liberated, but Austria, Czechoslovakia, and Norway continued under Nazi puppet governments that were under the control of Berlin. Nazi fascism dominated central Europe. For the next two decades, leaders in the United States and Britain were able to control the antidemocracy regimes in Nazi-controlled central Europe and in Soviet-controlled Eastern Europe by exploiting the hatred that existed between the Bolsheviks and the fascists. It was an uncomfortable peace, one that Winston Churchill described as an "abomination." The tensions that had exploded in 1939 in Europe continued in a cold war for the next several decades.

In the summer of 1940, the Japanese and Soviets signed a nonaggression pact. At a farewell party for the Japanese foreign minister, Stalin hugged his guest and said, "Now that Japan and Russia have fixed their problems, Japan can straighten out the Far East; Russia and Germany will handle Europe. Later together all of them will deal with America."

Adolf Hitler didn't agree with Stalin's plan. In the spring of 1941 he launched Operation Barbarossa, the invasion of the Soviet Union. Hitler had observed the difficulty the Soviets had in overcoming Finland, and he considered the Soviet Union a great house with decayed timbers. "We only have to kick in the door, and the whole rotten structure will come crashing down," Hitler said. The United States warned the Soviet Union that the Nazis were about to attack, but Stalin ignored the warnings. The Soviet Union had been shipping oil and other materials to Germany as recently as a week before the German invasion. Hitler had planned to launch his attack in the early spring and to wrap up the campaign in a few months—instead, the invasion of Soviet Union dragged on into the harsh Russian winter, when the Soviets were able to launch a major counteroffensive. Instead of quickly defeating Stalin and turning his attention to the planned invasion of England, Hitler now found that he was fighting a war on two fronts.

Victory in this two-front war was possible, Hitler believed, as long as the United States stayed out of the war, and he urged his Japanese partners to avoid doing anything that might provoke American into joining the conflict. Hitler told Japanese diplomat Hiroshi Oshima, "How one defeats the United States, I do not know."

Before the United States entered the war, U.S. military planners had drawn up battle strategies in case they needed to fight any enemy—including Britain. The strategies were named Red, Blue, Orange, etc., with the possible alliances between potential enemies labeled "Rainbow" plans. The eventual collaboration of the United States, Britain, and the Soviet Union against Germany and Japan was Rainbow Five.

The shuffling of the potential allies in World War II didn't end with America's entry into the war in December 1941. Just a few short days after Nazi Germany surrendered to the Allied forces in May 1945, Winston Churchill ordered plans for "Operation Unthinkable," for a joint U.S.–British invasion of the Soviet Union. The battle plan was delivered to Churchill on May 22, 1945, just two weeks after the Nazi surrender. The plan called for five hundred thousand British and American soldiers (47 divisions) to launch the attack through Germany, with the forces to be backed up by one hundred thousand *German* troops. The attack was scheduled to begin on July 1,1945. On June 29, 1945, the Soviet Red Army, which had stood down after the German surrender, suddenly redeployed their order of battle after apparently getting word of a possible attack. Obviously, the plan was never put into effect, but Britain and the United States would face the Soviet Union in an undeclared cold war for the next forty-four years.

Chapter 21

McAuliffe's Surrender

∞

In World War II's Battle of the Bulge, the largest battle ever fought by American soldiers, Brig. Gen. Anthony McAuliffe rejected German demands of surrender at Bastogne. What if he had accepted them?

IN 1944 the Allied forces sought to penetrate the continent of Europe, which German dictator Adolf Hitler had turned into a virtual fortress.

In the June 6, 1944, D-Day invasion, the Allies landed eight divisions—175,000 men. It was the largest invading force in world history, but the Nazi army was prepared to meet it with fifty-five divisions. Even if the landing was completely successful, the Allies would be outnumbered nearly seven to one.

Dwight Eisenhower, supreme commander of the Allied forces, chose the beaches in the French Normandy region, a poor place to make a landing, as the site of the invasion. Because the site was illogical, Nazi field marshall Irwin Rommel kept nine of his eleven armored divisions in Calais, France, north of Normandy. The landing at Normandy was a surprise, yet the German defenses there were still formidable. Some of the first Allied assault companies to land on Normandy's beaches suffered casualties of 90 percent. "Gentlemen, we are being killed on the beaches," American division commander Norman Cota told his men that day. "Let's go inland and be killed."

The Allied forces soon controlled the beachhead, and within days more than one million Allied troops were in France. By mid-August the Allies had recaptured Paris. (Hitler had ordered the retreating German army to burn the city to the ground, but the commanding general, Dietrich von Cholitz, not wanting to be known as the man who destroyed one of the world's most beautiful cities, disobeyed the order.) In October the Americans captured the first German city, Aachen, and the Allies had a solid line across Europe that was pressing its way into Germany. The line stretched from the Mediterranean Sea north to the southern border of Switzerland and picked up again at the northern border of Switzerland and continued to the Netherlands, running along the western border of Germany. It was a line of three-quarters of a million soldiers, mostly American and British, across the middle of Europe. "The enemy is at present fighting a defensive campaign on all fronts," said a confident intelligence report from Eisenhower's staff at the Supreme Headquarters of the Allied Expeditionary Force (SHAEF). "His situation is such that he cannot stage major offensive operations."

By conventional military strategy, Hitler should have pulled his troops back behind the Rhine River and defended Germany from that advantage. But Hitler would not stand for this insult to his military accomplishments. Eisenhower had achieved an early victory in the invasion of Europe with his surprise at Normandy—now Adolf Hitler had a surprise of his own. On December 15, the Germans launched twenty-four divisions against the American line in the Ardennes. At first the Americans couldn't believe what was happening. Gen. George Patton, who was famous for both his brilliant military maneuvers and his colossal ego, was commanding infantry hitting the southern flank of the German line. He told Bradley that the German attack in the Ardennes was a feint. "That's no major threat up there," he said. "Hell, it's probably nothing more than a spoiling attack to throw us off balance down here and make us stop this offensive."

But soon the German's eight elite Panzer Waffen SS tank divisions created a seventy-mile-wide gap in the line. American commanders rushed cooks, bakers, typists, and any other soldier to the front lines, but it wasn't enough to stop the German assault. One American division was completely destroyed, and many soldiers in nearby divisions were panicked.

In 1940, when the world had considered France's massive fortifications at the Maginot Line to be impenetrable, Hitler had been able to break through them by sending tanks through the Ardennes forests, which were lightly defended. The Ardennes region of Belgium and Luxembourg consists of heavily forested hills filled with bogs and soft ground. Tanks and other mechanized war machines were forced to travel through the Ardennes on the roads instead of through the fields, making the armor vulnerable to attack. The Nazi strategy of sending tanks through the Ardennes, which had worked so well in 1940, seemed to be about to work again.

The Nazi army had hit the American line with technologically advanced Tiger tanks and fresh troops that Allied intelligence reports had said didn't exist. "Pardon my French," cried Omar Bradley, the American commander in Europe. "But just where in hell has this son-of-a-bitch gotten all his strength?"

The answer was that the Nazis had forced young teenagers, fifteen and sixteen years old, into the front lines.

Bradley decided that he had to hold the "crucial road junction" Belgian town of Bastogne, and he ordered the 101st Airborne Division to go there to join the Ninth Armored Division. Bastogne was located near the Belgium-Luxembourg border some ninety miles southeast of Brussels. The twelve thousand men of the 101st Airborne rushed all night in open-air trucks through the bitter cold to reach the town. Unfortunately, the next day, the Germans surrounded the city with four infantry divisions and two armored divisions, trapping the Americans.

"Even though it might cost us heavy casualties in the airborne division and the two armored combat commands that had reached that outpost, I could not afford to relinquish Bastogne and let the enemy widen his Bulge," Bradley later wrote. Bradley told Patton to stop his offensive on the underbelly of the German line and to hurry to the aid of the men trapped at Bastogne. Bradley thought that getting there would take Patton four days—and that would be only if everything went smoothly. Patton said he could get his men to Bastogne in forty-eight hours. Patton pointed his cigar at the Bulge. "Brad, this time the Kraut's stuck his head in a meatgrinder," he said. "And this time I've got hold of the handle."

On December 22 four German soldiers, two officers and two young privates, came walking toward the American line in Bastogne with a white bedsheet tied to a pole. They told an American sergeant that they wanted to talk with the commanding general. The officers were blindfolded and taken to the headquarters of Brig. Gen. Anthony McAuliffe. McAuliffe was sleeping when one of his aides woke him and read the

German note:

> *To the U.S.A. commander of the encircled town of Bastogne,*
>   *The fortune of war is changing. This time strong German armored units have encircled the U.S.A. forces in and near Bastogne.*
>   *There is only one possibility to save the encircled U.S.A. troops from total annihilation: that is the honorable surrender of the encircled town.*
>   *If this proposal should be rejected one German Artillery Corps and six heavy A.A. Battalions are ready to annihilate the U.S.A. troops in and near Bastogne.*
>   *All the serious civilian losses caused by this artillery fire would not correspond with the well-known American humanity.*
> *The German Commander*

McAuliffe knew that he was surrounded and that his men were nearly out of ammunition and food. Bullets had already been rationed at ten rounds per day per gun. He had four hundred wounded men and few medical supplies. A supply drop was scheduled for ten o'clock that evening, but the weather had been poor and no airplanes had been flying for days. The German letter was correct: The Americans' situation was bleak. But Eisenhower himself had ordered that the crossroads at Bastogne be held, and McAuliffe had no intention of giving up. He took the papers in his hand, looked at them, and said, "Nuts!"

McAuliffe left his headquarters in a jeep to drive out to the front without bothering to respond to the German demand. When he returned, the German officers insisted that they had delivered a formal military communication and deserved an answer. Although McAuliffe had made it clear what he thought of the German demand that he surrender Bastogne, he wasn't sure how he could convey his thoughts to the Nazi commanders. Finally one of his aides suggested that his original remark had best summed up the feelings of the Americans, and so McAuliffe wrote out a response:

> *To the German Commander:*
> *Nuts!*
> *The American Commander*

When McAuliffe's colonel passed the general's answer to the German officers, they hesitated. "But what does it mean?" one asked.

"In plain English, it's the same as 'Go to hell,'" the colonel said.

With that the Germans saluted sharply and left.

The Germans weren't the only ones confused by the response. The French press association later translated the note as, "You are nothing but old nuts." And many in the U.S. military didn't believe that "nuts" was the exact word McAuliffe had used when he read the surrender demand. ("Can you imagine the commanding general of the 101st Airborne saying anything but 'shit'?" asked Kurt Vonnegut, a WW II infantry scout and later author of *Slaughterhouse Five*.)

Despite the bravado, the situation in Bastogne was bleak. McAuliffe's commanders at the edges of the city were requesting artillery to drive back the German troops, but McAuliffe refused to allow it. "If you see four hundred Germans in a hundred-yard area, and they have their heads up, you can fire artillery at them," McAuliffe said. "But not more than two rounds."

A small force of three tanks and two half-track trucks was organized to try to break through the German lines and retrieve ammunition. But before the squad had gone one mile, all five vehicles were destroyed.

By midnight, it became clear to the Americans that the promised supply drop wasn't coming. The weather was too severe for the airplanes to fly. Then unfamiliar airplanes could be heard overhead, followed by explosions. The Americans hadn't risked a rescue flight, but the Germans had sent bombers to destroy Bastogne. McAuliffe's flip reply had been forwarded up the chain of command all the way to Adolf Hitler, who demanded that the town be taken immediately, no matter how heavy the German losses.

As the Germans rerouted men and equipment needed elsewhere to try to crush the American army at Bastogne, the men trapped in the city faced a bleak Christmas. McAuliffe sent a mimeographed message to the soldiers on December 25th: "What's merry about this, you ask? We're fighting—it's cold—we aren't home. All true, but what has the proud Eagle Division accomplished with its worthy comrades? . . . We continue to hold Bastogne. By holding Bastogne we assure the success of the Allied armies."

On that Christmas Day, Patton had pushed his Third Army, 133,000 tanks, trucks, and pieces of machinery over seventy-five miles of ice- and snow-covered roads in just two days. Bradley called Patton's pivot "one of the most astonishing feats of generalship of our campaign." Finally, on December 26, clear weather allowed the American to put their airplanes into the skies. Two thousand bombers accompanied by 800 fighter planes attacked German positions, while 241 troop-carrier aircraft made low-level drops of food, ammunition, and medical supplies to the troops at Bastogne. Then, on the afternoon of December 26, Patton's Fourth Armored Division punched through the ring around Bastogne and ended McAuliffe's week-long holdout. The Americans had lost nearly five hundred men in Bastogne, with another twenty-five hundred wounded, but the critical crossroads in the center of the Bulge did not fall to the Germans.

WHAT IF?

What if General Anthony McAuliffe had agreed to surrender the crossroads at Bastogne?

Brig. Gen. Anthony McAuliffe stood in the basement headquarters of the 101st Airborne, staring at the German note. "Nuts!" he said, throwing the piece of paper to the floor. He marched up the steep steps of the building and left the headquarters, going out to one of the artillery units on the edge of town to personally congratulate the soldiers on fending off a German unit the day before. The men reminded McAuliffe of something that he was well aware of already—they were almost out of ammunition. Some were full of bravado and bile when McAuliffe asked if they were ready for the next German attack, but he could see that several dropped their heads and didn't answer. About a third of the men had had enough, McAuliffe figured, and he knew that as the ammunition disappeared, so would the men's resolve.

McAuliffe returned to the basement headquarters and was reminded that the German soldiers expected an answer to their surrender demand. McAuliffe turned to one of his staff. "What should I say?"

The major laughed. "What you said earlier was pretty good. Why don't you just say that?"

McAuliffe smiled briefly, then shook his head. "No, we need to do this thing. We have to . . . do this. I . . ."

McAuliffe paused. "They must . . ." He stopped again, trying to collect himself. He sat down and wrote out the acceptance of the surrender himself.

With the capture of the crossroads at Bastogne, the Germans were able to move more divisions north. The Allies assumed that the Sixth Panzer Army would try a direct line to the bridges across the Meuse River, but instead they drove toward Spa, where the Americans had a major fuel depot. On December 28, the Germans captured nearly four million gallons of much needed fuel.

To the amazement of the Allied foot soldiers, the German troops seemed to gather strength as they pushed west. This was not just the melancholy of Allied troops on the retreat: There were several reasons why the German army did in fact become stronger as it pushed farther from its line. In the early morning hours of December 15, as the German counterthrust began, the German army had dropped thousands of paratroopers behind the Allied lines. As had happened in the Allied D-Day invasion, nearly all of the paratroopers had been scattered far from their targets, and as the German line pushed westward into Belgium, these German soldiers rejoined their ranks.

The German commander, Field Marshall Gerd von Rundstedt, had gambled by sending his mechanized army into the Allied lines. He lacked the basic supplies, food, ammunition, and fuel needed to achieve his goal. Consequently the German panzer divisions, rather than heading toward the Allied forces, raced to capture supplies at every Allied depot.

The Sixth Panzer Army crossed the Meuse River at Liege and then turned north toward Antwerp. The British had captured Antwerp, Europe's largest port, in Septem-

ber of 1944 and immediately made use of the docks and
warehouses there to store supplies. The panzers raced
toward them while the German Seventh Army fought
Patton's Third Army in the south and the German Fifteenth Army battled the British Twenty-first Army Group
to the north.

As British field marshall Bernard Montgomery
shifted divisions to the west and south to try to hold
Antwerp, Germany's Fifteenth Army captured the Allied
communication center at Maastricht, Belgium. With the
Allies unable to communicate and their supplies under
threat, the Allied invasion, which had begun so successfully in the warm summer months, began to unravel. The
Allied headquarters issued orders to Bradley that he was
to try to slow the German advance in any way that he
could. "What the devil do they think we're doing?"
Bradley asked. "Running back toward the beaches?"

In late January the assault on Antwerp began. From
the first days of the German offensive in the Battle of the
Bulge, the overriding fear of the Allies would be that the
Germans might divide the Allied forces, separating
Bradley's army in the south from Montgomery's army in
the north. If that were to happen, Montgomery's British
and American troops would be trapped against the sea,
just as the British army had been at Dunkirk at the beginning of the war. To prevent another Dunkirk, Eisenhower
ordered Montgomery to shift his divisions to the south to
rejoin Bradley's army, so that the Allies could mount a
unified counteroffensive. Despite the consolidation of
forces, Antwerp fell to the Germans on March 2.

The German success in the Battle of the Bulge
prompted the British and American commanders to argue
over who was to blame. After Montgomery lost Antwerp

to the Germans, Patton tried to convince Bradley, "Let me go to Brussels and drive Montgomery into the sea, and then we can knock out the Germans without his screwing everything up." Bradley wrote later in his memoirs that he never did decide whether or not Patton had been joking.

In mid-March, 1945, just as it appeared plausible that the Allies might be forced off the Continent, as the British had been in 1940, the German thrust slowed. Like a strong wave on the beach that has expended its inertia, the German line began to pull back. On the eastern front the Soviet Red Army had reached the German border, and Hitler had ordered troops and supplies be redirected there as quickly as possible. Because of the captured Allied supplies, and the psychological boost provided by the success of the thrust through the Ardennes, the Germans were able to check the Soviet advance at the original German border. Bradley and Montgomery began to move through Belgium again, slowed only by the occasional attack by the Nazi's latest weapon, the "robot bomb," the world's first cruise missile.

In April 1945, Franklin Roosevelt died, just as the American forces were crossing the Meuse River in Belgium once again. Nazi propaganda minister Joseph Goebbels telephoned Hitler with the news, telling him, "My Fuhrer, I congratulate you! Roosevelt is dead! It is written in the stars that this will be a turning point for us!" For a time it appeared that Goebbels's astrology would be validated. The Allied march back toward Germany bogged down until it became a standoff, which reminded many of the last years of trench warfare in World War I. By June 6, 1945, a year after the D-Day invasion, the Allied line again reached only to the German border. Dresden and Hamburg were fire-bombed, just as Tokyo was. In both Germany and Japan, the United States had begun a strategy of trying to destroy the all of the major urban areas before invading the countries.

"Now that Hitler has been forced back into his cage, it is time for the United States to consider an honorable end to the war," an editorial in the *Nation* advised President Harry Truman. A small but growing minority of people in the United States agreed that it was time to end the war and to stop the bombing of civilians; others doubted that the costs of thousands more American lives justified the invasion of the two offending nations.

However, on July 16, 1945, the world was changed forever, and almost no one knew it. On that day, the United States conducted the first successful test of the atomic bomb. Three weeks later, an American B-29 dropped a single uranium bomb on the Japanese city of Hiroshima, destroying it completely. Three days after that, a second B-29 dropped a more powerful plutonium bomb on Berlin. Adolf Hitler was among those killed in the bombing: Hitler's residence had been the bull's-eye of the attack. The next morning Berlin had collapsed into rubble, with more than sixty thousand dead. The new American weapon was almost beyond belief—one bomb could destroy an entire city. The shock and power of the weapon was such that Tony McAuliffe even heard rumors of it while in a German P.O.W. camp. "I wondered, what is this Adam bomb? How could anything be this horrible?" he said later. Within days, both the Japanese and Germans surrendered, and World War II was over.

Even after Patton punched through the German ring around Bastogne and rescued McAuliffe's troops, Eisenhower's staff at the strategic headquarters told Bradley that they expected the Germans to be across the Meuse River within forty-eight hours. "Nuts!" Bradley yelled, intentionally plagiarizing McAuliffe's answer. Despite the pessimism of the planners at Supreme Headquarters, the Battle of the Bulge was over, and Bradley knew it. In the battle, losses on both sides had been severe: the Germans saw 30,000 men killed, 40,000 wounded, and 40,000 captured as prisoners of war; the Americans saw 20,000 men killed, 40,000 wounded, and 20,000 taken prisoner.

The German forces had pushed to within four miles of the Meuse but then began to fall back. By the end of January, 1945, the Allied line was back to the German border. More important, Hitler had used up the last of his troops and his military hardware in the offensive, sealing his doom. He had nothing left to fight a defensive battle at the Rhine, and within one hundred days Hitler would commit suicide and the reign of the Third Reich would be over.

After the war, Anthony McAuliffe remained in the army and supervised the ground forces at the atomic bomb tests on Bikini Island. As for the men of the 101st Airborne who fought at Bastogne, they were awarded the Distinguished Unit Citation for their role in holding on to the crossroads and spoiling the Nazi offensive. It was the first time in U.S. history that an entire division received the award.

Chapter 22

Dewey Defeats Truman

✧

Thomas Dewey, with a commanding lead in the polls following the 1948 political conventions, decided to coast to the
election, but Harry Truman was running for all he was worth.
What if Dewey had decided to run a vigorous campaign, too?

IT seemed as though it was over before it even began.

In the election of 1948, no one gave the incumbent president, Harry Truman, a chance against Republican challenger
Thomas Dewey—at times, it seemed, not even Truman himself. On July 31, 1948, both the Democrat Truman and
Republican Dewey attended the dedication of the Idlewild
Airport in New York. (Truman was big on the political benefit
of dedications—he boasted that he had dedicated the Grand
Cooley Dam three times.) During the event, Truman leaned
over and whispered in Dewey's ear that when he moved into
the White House, he would need to repair the plumbing.

Even Bess Truman, Harry's wife, had a hard time envisioning a Truman triumph. That summer she laughed to a
White House aide, "Does he really think he can win?"

✧

Harry Truman was like the poor soul who is hired to replace the nationally famous coach with a decades-long winning record. After Franklin D. Roosevelt had served as president for twelve years, guiding the country through the Great Depression and a world war, nearly everyone in the nation had a difficult time comprehending that he wasn't president any longer. Truman once mentioned a presidential appointment to an aide, who asked if the president had made the appointment before he died. "No," an annoyed Truman said. "He made it about a half hour ago."

Truman's task of winning the presidency on his own merits became even more difficult when Roosevelt's wartime vice president, Henry Wallace, broke from the Democratic Party to run for president on the Progressive Party ticket. This was sure to drain votes from the left wing of the Democratic Party. H. L. Mencken was dismissive of the Wallace movement, saying that it was an assemblage of "grocery-store economists, mooney professors in one-building universities, editors of papers with no visible circulation, preachers of lost evangels, and customers of a hundred schemes to cure all the sins of the world." But at the time, many considered Wallace's candidacy a real threat to Truman. In February 1948, there had been a special election for a congressional seat in the Bronx, an election that the Democrats had pointed to as an early test of public opinion, and the Wallace-endorsed candidate trounced the Democratic candidate.

The infighting in the Democratic Party continued at the Democratic national convention in Philadelphia in June. The party leaders had asked a young Hubert Humphrey to speak, and his rousing address in favor of civil rights propelled the party to add a civil rights plank to its platform. When this happened, thirty-five Southern delegates walked out of the convention. The Southern Democrats formed the States' Rights

Party, more commonly known as the Dixiecrat Party, and nominated Strom Thurmond as their presidential candidate. With the Democratic Party split three ways, it appeared that Truman had no chance of winning in the fall. "A stranger convention there never was," reported Lowell Mellett. "Man and boy, but I've never seen . . . one in which so many delegates worked so hard and at such cross-purposes with seemingly but a single intent—how to make sure that the man nominated should not be elected." The *New York Post* agreed, saying, "The party might as well immediately concede the election to Dewey and save the wear and tear of campaigning."

Thomas Dewey, the governor of New York, had been the Republican's nominee in '44, when he gave Franklin Roosevelt the closest of his four elections. Political wisdom said that a party shouldn't renominate a failed presidential candidate, but Dewey was able to overcome this belief and receive the party's nomination again in 1948.

Dewey was a short man, five-eight and dogged by rumors of elevator shoes. His angular cheekbones and chin prompted comments that his face looked as though it were squeezed by a vise. Dewey had first come to national attention in the early 1930s, when as a prosecutor, he courageously stood up to crime bosses such as Lucky Luciano and Dutch Schultz. His earnest do-gooder attitude had its detractors—one New York critic dismissed Dewey as "an honest cop with the mind of an honest cop"—but by the time he was thirty-five years old Dewey was a national celebrity.

Dewey wore his celebrity uncomfortably. He was often uneasy around people, and he was also something of a technocrat. His only outside interests existed on a dairy farm that he kept as a hobby, where he was fascinated by the technology of artificial insemination. He was known to dominate dinner conversations at the farm extolling the virtues of procreating via pipette.

As a politician, Dewey saw himself as a more modern politician than his opponents, one who believed in making politics a science instead of an art. He was the first presidential candidate to have his own staff to conduct public-opinion polls, and he put enormous faith in what the poll numbers advised him to do. By the summer of 1948, the polls showed Dewey that he had a huge lead over Truman. He believed that all he had to do was to play it safe, not make any mistakes, and the election would be his. "My job is to prevent anything from rocking the boat," he said in July.

Dewey preferred an aseptic form of politicking anyway—unlike Truman, he wasn't one to get out in the crowds and slap people on the back. When it came to inserting his ideas into the political arena, Dewey was a strong believer in artificial insemination.

Dewey once shared a convertible with a popular county chairman in Ohio, George Bender. As the car passed entire groups of people who waved and shouted, "Hello, George!" Dewey turned to Bender and asked, "Who the hell is this guy George?" At another campaign stop, the local VIPs overheard Dewey asking his aides, "When the hell do we get out of this damn town?"

On the campaign trail, Dewey kept a railroad car full of speechwriters in the train's next-to-last car, known as the "squirrel cage." When the speechwriters brought Dewey their speeches, he derided their efforts, typically demanding a dozen or more drafts of a speech before he found it acceptable (the record was said to be twenty-one drafts). When one speechwriter protested that one speech had already been rewritten many times, Dewey responded coolly that "practice is the difference between the amateur and the professional."

Despite all of the professional help, the words that came out of Dewey's mouth were breathtaking in their inability to take one's breath away. On the nation's business, he boldly asserted, "Government must help industry and industry must cooperate with government." On America's outlook: "Your future lies ahead of you." On the fate of nations: "The ancient civilization of Babylon is now a dead thing." To this incisive analysis he added, "This tragedy must never happen to America." Making things worse, Dewey liked to add the rhetorical flourish "Period!" at the end of sentences he wanted to emphasize, and his delivery was so stiff that *The New Yorker* said that at campaign rallies, he arrived "like a man who has been mounted on casters and given a tremendous shove from behind."

Truman poked fun at Dewey's cautious statements, saying that Dewey was running a "soothing-syrup campaign" and that now the GOP stood for "Grand Old Platitudes." If Truman was laughing, some Republicans were annoyed about the oatmeal being served by the Dewey campaign. Dewey's running mate, Earl Warren, grumbled, "I wish I could call someone a son-of-a-bitch."

Dewey's sterile, analytical style inspired few true believers, but it was his arrogance that irritated would-be supporters. A common quip among those who worked with him was that he was the only man alive who could strut sitting down. While he was prosecutor in New York, Dewey's arrogance had angered so many journalists that several photographers took photos of Dewey only in unflattering poses. *Time* magazine reported that those opposed to Dewey said that he was "too mechanistically precise to be liked ... too coldly ambitious to be loved." Dorothy Thompson reported that Dewey "had fewer real friends than any other leading candidate and that if he had been defeated more people would have been pleased with themselves."

Despite the shallow support for Dewey, it seemed that the election was a lock. "Dewey-Warren will be unbeatable. So: It's to be Thomas Edmund Dewey in the White House on January 20, with Earl Warren as backstop in event of any accident during years just ahead," said *U.S. News & World Report* in July. "Fifty political experts unanimously predict a Dewey victory," *Newsweek* reported in October. "Dewey is going to be the next president, and you might as well get used to him," *The New Republic* complained in October. In the fall, *Life* magazine put Dewey on the cover, with the caption "Our next President."

At the Republican national headquarters, a sign was posted to help campaign workers maintain their focus. "1,460 more days until election," it said, referring to date in 1952 when Dewey would face reelection.

The situation had become so desperate for the Democrats that they resorted to sports analogies. "We've got our backs on our own one-yard line with a minute to play," said Clark Clifford, neglecting to mention what down they were on or the field condition. "It's got to be razzle-dazzle." The desperation play that Truman used was to cast himself as an outsider, attacking those in Washington, especially the Republican Congress. It was an odd sight, the Democrat Truman, who had served in Washington for decades and whose party had been in control of the White House for sixteen years, attacking New York governor Dewey and the newly installed Republican majority for their Washington ways. In September, a *Newsweek* cover caption asked, "Who is Challenger, Champ?"

But for those who cared to notice, there were signs of trouble in the heartland for the Dewey campaign. The Staley Milling Company offered farmers a choice between feed bags with a donkey or an elephant stenciled on them, and when it was obvious after twenty thousand bags of feed that the donkey was pulling away, the company decided to stop the burlap poll. In another attempt to discern the thinking of the working man, in August a Truman crony disguised himself as a chicken farmer and drove an old truck along the Ohio River. After three weeks he reported back to the White House that Truman was going to win.

One cause of Dewey's eroding support was falling crop prices. At the beginning of 1948, corn prices were $2.50 a bushel; by September the price had dropped to $1.78, and it would eventually fall to less than a dollar a bushel by election day. On his whistle-stop train ride across America, Truman was quick to point out that the New York lawyer wasn't sympathetic to the hard life of farmers, saying that Dewey planned to "stick a pitchfork in every farmer's back" by cutting government price supports enacted during Roosevelt's New Deal. Truman, of course, also attacked the Republican Congress (always a convenient punching bag), and at times he even took the low road, telling audiences that the uptight Dewey was "a front man" for the people who had backed fascists such as Hitler, Mussolini, and Tojo.

That caricature of Dewey stuck after an incident on the Dewey campaign train Victory Special in Illinois. At a campaign stop, the train had backed up a few feet instead of moving forward, almost crushing those in the crowd who had gathered to hear Dewey, including one of his closest friends. "That's the first lunatic I've had for an engineer," Dewey snapped at the microphone. "He probably ought to be shot at sunrise." Organized labor used the Dewey quote to show that the Republican candidate didn't care for the working man, and newspapers around the country printed the train engineer's response: "I think as much of Dewey as I did before, and that's not very much."

Truman's campaign had gained traction, but almost no one in the country knew to what extent. Many pollsters had stopped taking polls in September because they considered Dewey's lead to be insurmountable. Still, some Republicans were worried, and they pleaded with Dewey to go out and campaign more vigorously. But Dewey refused. "I waged an all-out fight in 1944 and lost," he said. "In 1942 and 1946 I waged a different kind of campaign for the [New York] Governorship, not even mentioning my opponents, and won by large margins." He continued his cautious ways: At a rally in the Los Angeles Coliseum where ninety thousand supporters showed up, after the crowd was pumped up by a parade of Hollywood celebrities, Dewey took the stage and began an economics lecture on the social security system. "This is wrong, all wrong," one Dewey aide worried.

The night of the election, Dewey supporters gathered in the lobby and in the hallways of the Roosevelt Hotel in New York, the same hotel where Dewey was awaiting the election returns, to celebrate and drink champagne. In the early morning hours, while the revelers enjoyed the presumed victory in the hallways, inside Suite 1527 there were white faces and hushed conversations. Thomas Dewey sat quietly in the bedroom with a yellow legal pad in his lap, carefully recording his political demise.

As the early election returns had begun coming in from the East Coast, Truman was leading, but this was hardly unexpected. Many Democratic races had gone this way, with the tide turning toward the Republicans as the returns began flowing in from the Midwest and mountain states. But the returns from the heartland were making the night of his triumph a nightmare for Dewey. Even Iowa, which a senator had said would "go Democratic the year hell goes Methodist," went for Truman.

At three in the morning, George Gallup told a reporter that it was likely that his polls had been wrong and that Truman was elected. The reporter wrote that Gallup looked like an animal forced to eat its young.

In Chicago, the night editor faced a difficult decision. The article announcing Dewey's victory had been finished for two hours, but as the time to go to press neared, Truman still held a half-million-vote lead. Finally the editor decided to go with the now famously wrong banner headline: DEWEY DEFEATS TRUMAN.

WHAT IF?

What if Dewey had not relied on his scientific polls and aseptic campaigning, and had waged a vigorous campaign against Truman in the 1948 election?

The reaction to his campaign-stop comment about the train engineer in Illinois had startled Dewey. It was an impulsive comment made in a flash of anger, but now people around the country were using the remark as a peg on which to hang a new portrait of him. The sense of the attacks—that he was unfriendly to labor—didn't bother Dewey as much as the erosion of support the attacks were causing. After the train engineer debacle, he had ordered a new round of secret polls, and the numbers were stunning. In many areas of the country, he was actually running even with Truman now, and in some rural areas he was falling behind!

Dewey decided that he had to mount a more aggressive campaign, but the question was how. He was offering himself as a more efficient and responsible candidate for the presidency—on most issues, there wasn't any difference between his position and Truman's. The single significant difference was in international relations. Truman was in over his head when dealing with global strategy, Dewey thought, and too quick to accept the explanations offered by the Soviet Union for its actions. Truman had even referred to Stalin as "good old Joe." Dewey, the former prosecutor, considered Stalin just another mob boss, and thought that he alone was capable of containing him.

In late October, Dewey went on a vigorous attack of Truman's foreign policy. "I agree with President Truman's decision to airlift supplies to Berlin," he said. "But I don't think he understands the type of cold-blooded people he is dealing with. I have faced such men in the mobs of New York, men with many notches in their guns, and I won't agree to friendly terms with 'Uncle Joe.' I will increase defense spending by $25 billion, and I will improve our ties with those who agree with our policies, such as Charles de Gaulle and Chiang Kai-shek. We will only support those who support our interests abroad, especially our ancient friend and ally, China."

Dewey accused Truman of going too easy on the Soviet Union and the Democrats of having caved in to Soviet demands in the conferences at the end of World War II at Yalta, Quebec, Teheran, and Potsdam. Dewey believed strongly that the Berlin airlift was necessary only because corridors to Berlin had not been negotiated by Truman at Potsdam. "These failed negotiations have cost our nation $30 billion dollars, money that could have been better spent here at home," Dewey charged.

Dewey had miscalculated the degree of concern over international issues in the nation's heartland. Few people on Main Street cared about geopolitical issues, but the point was moot. The farmers and mechanics may not have spent much time thinking about international chess games, but they held a strong band-playing, flag-waving Fourth of July patriotism, and Dewey's calls for increased military spending and offering support abroad only to those governments who looked out for American interests was a perfect fit with their barbershop philosophies.

He also had strong words for the right wing of his own party: "Many are rightfully disturbed that Franklin Roosevelt, who has been in the grave for three years, still defines the public debate over our policies. But I will not abandon programs such as Social Security, unemployment insurance, and farm supports. Such programs cost us very little when compared with the gain in human happiness and security they provide. We need a pragmatic liberalism, of the type initiated by Republicans such as Abraham Lincoln and Theodore Roosevelt, that provides private incentive while maintaining the public conscience."

A week before the election, Dewey's aides told reporters that he was going to make a major announcement about farm policy at his dairy farm in New York. The statement was insignificant (a repeated vow to continue farm subsidies), but Dewey appeared in work clothes and allowed photographers to snap photos of him milking cows and cleaning out the stalls. Dewey picked up a shovel full of manure (a cow patty had been carefully positioned and its location mapped out in the briefing), smiled, and said, "I'll have to clear out a lot of this in Washington." The next day, the photo and the quip ran on the front page of many newspapers across the country.

Many people, including Dewey himself, considered the dairy farm press event to be vulgar and inappropriate for a presidential candidate. But Dewey was pleased to see that within days, his own polls showed that he had pulled even with Truman among farmers.

On November 2, 1948, the election was as lopsided as the midsummer polls had predicted. Dewey won 55 percent of the popular vote, Truman received 41 percent, and Thurmond and Wallace received 2 percent each.

Almost immediately after being inaugurated on January 20, 1949, Dewey had to turn his attention to the civil war in China. Mao Tse-Tung's communist rebels were driving Generalissimo Chiang Kai-Shek's Nationalist soldiers to the south, and Chiang was in danger of seeing his government overrun. Within days after becoming president, Dewey signed legislation giving $2 billion in loans to Chiang's government. Dewey also offered the use of U.S. ships and airplanes to transport Nationalist soldiers and announced that he was sending sixty thousand soldiers to China to act as military advisors.

Dwight Eisenhower, now president of Columbia University, spoke out against Dewey's China policy. "Any U.S. involvement in the internal politics of the nations of Asia is bound to become a national tragedy," he said. Democrats and Republican isolationists also questioned why America needed to send troops to the world's most populous nation. Dewey continued to commit more American "military advisors," however, until by the end of

1949 more than one hundred thousand American soldiers were fighting in China.

Just as the voices of Democratic critics of the war in China were starting to have an effect on Dewey's polls in the heartland, Dewey had to contend with trouble from within his own party. In the autumn of 1949, a young congressman from California, Richard Nixon, began talking to reporters about Communist spies in the State Department. Dewey was furious over Nixon's comments, and he telephoned Nixon himself and explained to the young congressman what an embarrassment he was causing the new administration. "If this continues, I guarantee that you will face a strong challenge in the next primary, one that you will lose," Dewey said. Nixon decided to leave the business of uncovering spies from the House Committee on Un-American Activities to the Justice Department, and Dewey went to California to campaign for Nixon in his run for the Senate in 1950. (After two terms in the Senate, Nixon retired from politics following a humiliating defeat in the 1962 California gubernatorial election.)

More significant than the congressional commie-chasers (Dewey said that he had to "squash Joe McCarthy like a cockroach") was the problem of the failing war in China.

Initially, in the summer of 1949, after the arrival of the United States' aid and weapons, Chiang was able to push the Communist troops back to the north and prevent them from taking Peking. But by late 1950, the effort wasn't going as well as the amount of aid that the United States was pouring in indicated that it should. Not only were Mao Tse-Tung's guerrilla war techniques effective, but Mao also had the support of the Chinese people. Chiang's government had treated the peasants as chattel or cannon fodder and denied even the soldiers such basics as uniforms (which the United States had paid for) and food (which the United States had shipped over). The Chiang government was so corrupt that even his own officers were selling the American weapons to the Communist rebels.

Douglas MacArthur, frustrated at the ineffectiveness of the bombing missions, began saying, both privately to Dewey and in public interviews, that the United States should use the atomic bomb in China. Dewey and Mao Tse-tung were in agreement that the atomic bomb would have little influence in China. Mao's guerrilla techniques didn't rely on large military bases or urban centers; most of Mao's soldiers were in the vast rural areas of China.

Instead, Dewey responded by sending additional American troops to China, increasing the number to more than two hundred thousand. Despite the infusion of American soldiers, Chiang still was slowly losing ground to Mao's rebels. By the summer of 1950, even Dewey thought that the chances of a Nationalist Chinese victory were only fifty-fifty. "The only way we're going to win is to get rid of Generalissimo Chiang," Dewey complained in a cabinet meeting. When the CIA suggested that it was available to "correct this situation," the straight arrow Dewey blanched. The president decided that America was doing as much as it could and that no further assistance, financial, military, or covert, would be going to China.

As the war continued into 1952, and the number of American soldiers killed in what the ex-prosecutor Dewey termed a "police action" increased, public opinion of the war fell. As Dewey began preparations for his reelection in the spring of 1952, he and the rest of the nation were surprised when Dwight Eisenhower announced that he would enter the primaries as a Democratic candidate for president. That fall Eisenhower easily defeated the unpopular Republican Dewey, and Eisenhower immediately began working on a partial withdrawal of American troops from Asia.

In May 1953, the United Nations was able to secure a cease-fire between the Communists and Nationalists. The Communists controlled Mongolia and Inner Mongolia; the Nationalists controlled Peking and all of China to the south. The two countries established a heavily armed border that, in places, incorporated the Great Wall of China, and the Nationalist Chinese were able to prevent the Communists to the north from spilling over the border only because of a large and apparently permanent U.S. military presence along the border. There were now two Chinas, just as there were two Koreas, two Vietnams and two Germanys.

Containing Communism, a job that the arrogant Dewey had thought Truman was unable to perform, turned out to be more nuanced than Dewey had realized. His eagerness to support a corrupt and inept regime in China left Dewey with what many considered a failed one-term presidency.

Truman won the election of 1948, but at times he must have wondered what he had worked so hard for. In December 1949, Chiang Kai-shek and his followers fled to the Chinese island of Taiwan and established their government there. China was split, but because the Nationalists controlled only a small portion of the great nation, for all practical purposes China was in complete control of the Communists. Republicans immediately charged that the Truman administration had "lost" China, and two months after the fall of Chiang's government, Wisconsin senator Joseph R. McCarthy announced that he had a list of 205 State Department officials who were working for the Communists. California representative Richard Nixon had a much shorter list, naming only Alger Hiss as a Communist, but it turned out that that was still one more name than McCarthy actually had.

In June 1950, North Korean Communists, buoyed by Mao's victory, invaded South Korea. Truman began sending troops and funds to both Taiwan and South Korea. By the end of the Korean War, 1.8 million Americans had served in Korea, with sixty-five thousand soldiers killed or missing in action. The public became frustrated with the difficulties in Asia, and in 1952, Dwight Eisenhower, running as a Republican, defeated Democrat Adlai Stevenson for the presidency with a promise to go to Korea himself to help bring an end to the war.

For his part, Dewey never accepted that his inability to connect with the public caused his loss to Truman. He believed that he had lost merely because Americans hate seeing a candidate run away with an election. "I have learned from bitter experience that Americans somehow regard a political campaign as a sporting event," Dewey told Eisenhower.

Chapter 23

Scrapping the U-2

❧

The CIA proposed building a secret spy plane that could fly over foreign countries and bring back photographic intelligence. What if President Dwight Eisenhower had considered the U-2 plan too risky?

ON May 3, 1960, the National Aeronautics and Space Administration released a statement:

A NASA U-2 research airplane, being flown in Turkey on a joint NASA-USAF Air Weather Service mission, apparently went down in the Lake Van, Turkey, area about 9:00 A.M. (3:00 A.M. E.D.T.), Sunday, May 1.

During the flight in southeast Turkey, the pilot reported over the emergency frequency that he was experiencing oxygen difficulties. The flight originated in Adana with a mission to obtain data on clear air turbulence.

A search is now under way in the Lake Van area.

The pilot is an employee of Lockheed Aircraft under contract to NASA.

The U-2 program was initiated by NASA in 1956 as a method of making high-altitude weather studies.

The statement was a bit inaccurate.

In truth, a Soviet missile had shot down a U-2 spy plane carrying a CIA pilot, Francis Gary Powers, as Powers took aerial spy photographs of military installations in the Soviet Union. Although the Soviet Union immediately announced that it had shot down the spy plane, U.S. officials in the White House and the CIA believed that there was no chance that the pilot survived, and so the Soviet Union would not be able to offer definitive proof to the world of U.S. spying. The CIA had taken elaborate precautions to prevent a U-2 spy plane or a CIA pilot from being captured by the Soviets, and they had insisted to President Dwight Eisenhower that it could never happen. When Soviet premier Nikita Krushchev announced to the world that the very-much-alive Powers was sitting in a Soviet jail, the U-2 incident became the United States' largest diplomatic embarrassment of the twentieth century.

"The CIA promised us that the Russians would never get a U-2 pilot alive," complained an angry John Eisenhower, a member of his father's White House staff, "and then they gave the son-of-a-bitch a parachute!"

❧

In 1953, the Soviet Union tested a hydrogen bomb, and the implications detonated in the United States. Now both superpower nations had the ultimate weapon, and the result was a fearful standoff. Two years later, just after Dwight Eisenhower took the oath of office as president of the United States, the Soviets unveiled a massive intercontinental bomber, which U.S. intelligence officials nicknamed the Bison. No one thought that the Soviets had developed this huge, long-range airplane to deliver relief supplies around the world. The Bison had been created to drop a nuclear bomb on a target in the United States.

The Soviet Union now had the capability to deliver what Eisenhower called "a second Pearl Harbor," and the chance of a surprise attack became the president's greatest fear. The only difference would be that, unlike Pearl Harbor, this time the surprise attack would be against civilians and would use nuclear weapons. Eisenhower needed to find a way to stop such an attack, or at least be warned that it was coming, and so he appointed a commission, headed by the president of MIT, to study the problem.

As a closed society, the Soviet Union held a distinct advantage over the United States when it came to spying. In the Soviet Union, even the Moscow telephone book was a classified document; in the United States anyone could move about the country and buy detailed maps and reference books, or take any number of photographs of bridges, factories, and military installations. In the United States, espionage was easy; in the Soviet Union it was nearly impossible.

The commission recommended that the United States begin reconnaissance flights over the Soviet Union to gather photographic intelligence. Aerial reconnaissance was certainly not a new idea. In 1794 a French captain had flown tethered balloons over the enemy during the Battle of Fleurus, and during the American Civil War Lincoln had created the Army Aeronautic Corps, which attempted to use hot air balloons to gather information on the Confederate army. Eisenhower already understood the value of photographic intelligence. An analysis after World War II had shown that 80 percent of the useful military intelligence gathered during the war had come from photographs. Photographs were simply more accurate and reliable than human spies. Without photographic intelligence, Eisenhower could not be sure how many bombers the Soviet Union had. (And his doubts about human intelligence were justified. The Soviets skipped numbers on the airplane, jumping from thirty-six to forty for example, leading observers to assume that bombers numbered 37, 38, and 39 were in some other location when in fact they didn't exist.)

But low-flying airplanes had taken the photographic intelligence during World War II; no such flights could take place over the middle of the Soviet Union. The problem was that the technology for high-altitude spying didn't exist.

To fly safely above the range of Soviet surface-to-air missiles, a spy plane would need to fly at seventy thousand feet—twenty-five thousand feet higher than any plane could fly at the time—and it would have to maintain that height for a flight of ten hours. At such heights, the air was so thin that a jet engine would produce only one-twentieth the thrust that it would produce at sea level. Some Air Force officials advised the president that such a plane simply could not be built.

One airplane designer disagreed. Clarence "Kelly" Johnson of Lockheed, who had designed such famous military planes as the World War II-era P-38; the F-80, the first U.S. jet fighter; and the C-130 Hercules cargo airplane, had sketched out a plan for a high-flying spy plane. Johnson's design had been turned down by the Air Force because it didn't believe in building specialized aircraft. Unless a plane could carry weapons and had other capabilities, such as serving as a fighter plane, the Air Force refused to built it.

The CIA gave the project to built the spy plane to Johnson and his engineers at Lockheed. The plane the engineers came up with was little more than a titanium kite with a jet engine attached. The designers went to great lengths to make the plane as light as possible. The tail was attached to the fuse-lage with just three bolts. The airplane lacked landing gear—the wheel assemblies fell to the ground at takeoff, and the plane returned to Earth on skids in a modified belly-landing. In the U.S. military, planes are assigned names based on their functions, for example the F-86 for "fighter" or the B-52 for "bomber." The new plane couldn't be given an "S" for "spy plane," so instead it was labeled the U-2, with the "U" standing for "utility."

Other technologies were developed to go along with the U-2. Eastman Kodak developed film as thin as plastic wrap that could be used in large quantities. Edwin Land, the inventor of the Polaroid lens and the Polaroid camera, developed a camera that swept from side to side and could take stereo images, and a Harvard astronomer developed a lens for the camera that was able to take a legible photo of a newspaper headline from eight miles in the sky. When these new tech-nologies were loaded into the nose of the U-2, the plane was able to photograph a strip of land 750 miles wide and more than 2,000 miles long, the equivalent to a swath of the United States from Phoenix, Arizona, to Washington, D.C.

Before he approved the secret missions over the Soviet Union, Eisenhower made an offer to the Soviets. At the Geneva Summit Conference in 1955, Eisenhower proposed an international Open Skies policy and invited the Soviets to con-struct airfields in the United States. The Soviets would be free to fly anywhere in the United States (with a U.S. observer on board). And, of course, the United States would expect the same courtesy in the Soviet Union. Just as Eisenhower spoke his final word, there was a loud clap of thunder and the lights went out. The scene was an apt metaphor—Nikita Khrushchev turned down the Open Skies proposal, and the world was plunged into the winter darkness of the Cold War, the most bleak and dangerous years of that decades-long struggle.

Eisenhower was concerned that if he approved the spy flights over the Soviet Union, the Soviets might mistake a U-2 flyover as the first plane in an American attack, and so he insisted on approving each flight over Soviet territory personally. In several White House briefings, Eisenhower would pore over maps, listen to the rationale for the proposed flight, and then either approve the flight with a pat on someone's shoulder or, with a vigorous shake of his head, state that the possible gain wasn't worth the risk. "Such a decision is one of the most soul-searching questions to come before a president," Eisenhower said. "We've got to think about what our reaction would be if they were to do this to us."

Eisenhower's biggest concern was that a plane—or, even worse, a pilot—might be captured by the Soviets. "If one of these planes is shot down, this thing is going to be on my head," Ike said. "I'm going to catch hell. The world will be a mess."

The CIA assured the president that it would be impossible for the Soviets to capture the U-2. A hit by a surface-to-air missile would blow the airplane to bits—the plane was so fragile that on a test flight over Germany, two Canadian fighter planes buzzed in for a closer look, and the U-2 disintegrated from the shock waves caused by the fighters. The plane also carried a two-and-a-half-pound explosive charge behind the pilot's seat. If the pilot needed to eject, he was to flip a switch that would cause the plane to blow up seventy seconds later.

What's more, the U-2 pilots were also given a special half-dollar, which could be pulled apart. Inside was a pin, a minute syringe loaded with a powerful toxin that the CIA had developed from a sea crustacean at a reported cost of $3 million. The pilots weren't ordered to commit suicide, though. Allen Dulles, head of the CIA, thought it was more likely that a man would kill himself if the decision were left up to him, that there would be a greater chance of "a man's individual nobility prompting him to such an act."

With such precautions in place, Eisenhower approved the U-2 flights, and the first U-2 flight over the Soviet Union took place on July 4, 1956.

After the initial round of flights over the Soviet Union, to the surprise of the CIA, the government in Moscow sent a secret diplomatic protest. U.S. military installations had not been able to track the U-2 in test flights over the United States, but the Soviets were able to track the flights over Russia, although they were not able to shoot the planes down. Soviet radar was clearly more advanced than the Americans'.

From seventy thousand miles up, American pilots were able to look down on Soviet air bases, missile launching pads, submarine bases, and atomic testing grounds. By 1959, 90 percent of what the United States intelligence services knew about the Soviet Union's military capability came from U-2 flyovers.

Although the U-2 program was shrouded in secrecy, it involved hundreds if not thousands of people at various sites around the world. Information about the airplane and its purpose would inevitably soon come out. By late 1958, many newspapers in the United States had knowledge of the U-2 and the flights over the Soviet Union. Officials from the Eisenhower administration were able to convince the newspaper editors and publishers to withhold the stories to protect national security, but apparently not every publication got the message. In 1958, the magazine *Model Airplane News* reported that "U-2s are flying across the Iron Curtain taking aerial photographs."

By the spring of 1960, the U-2 program was winding down because of security concerns and worries about improved Soviet antiaircraft missile capability. Eisenhower and Khrushchev agreed to meet in Paris for a disarmament summit in mid-May. The CIA wanted to make one last mission over the Soviet Union—everyone remembered how the Japanese had been negotiating for peace up until the hour that they attacked at Pearl Harbor—but Eisenhower said that the last possible day had to be May 1. Anything after that stood a chance of wrecking the summit.

On May 1, 1960, a young lieutenant sat in an unmarked U-2 on a hot runway in Pakistan. Francis Gary "Frank" Powers was to fly 3,800 miles from Pakistan to Norway, crossing the Soviet Union at the Ural Mountains. While he was over the western Soviet Union near Sverdlovsk, Powers's U-2 was disabled by the close explosion of a Soviet missile. Powers explained later that he had been thrown from the cockpit by the explosion, attached to the airplane only by his oxygen hoses, and had been unable to reach back inside the airplane to hit the destruct switches. That may have been true, but it was also true that many of the U-2 pilots didn't believe the CIA when it said that there was a seventy-second delay on the explosives behind the pilot's seat. The history of espionage contains many accounts of spies being killed after their usefulness is spent, so that they can't tell their tales to the enemy, and many of the U-2 pilots believed that there might not be any delay on the explosives at all.

As Powers floated to Earth below the orange and white canopy of his parachute, he opened the silver dollar and held the pin between his finger and thumb. Then, his mind made up, he slipped the pin into his pocket.

Powers was captured by the Soviets, and on May 7, Khrushchev addressed the Supreme Soviet and presented photographs of the plane wreckage and of Powers, who, Khrushchev announced, was "alive and kicking."

At the Paris summit a week later, instead of working together to find ways to remove arms and achieve peace, both the Soviets and the Americans engaged in angry denunciations of the other. Khrushchev actually spat out a loud, obscene series of insults at Eisenhower, who sat unaffected because the Soviet translator cleaned up Khruschchev's language before it reached the president's ears. American staffers who spoke Russian, however, were horrified. The chance for peace that the Paris summit had represented was clearly past.

WHAT IF?

What if Dwight Eisenhower had decided that building the U-2 spy plane was too risky?

"I just can't get past the thought of what we would do if the Russians sent planes over Dallas or Chicago. What would we do?" Eisenhower looked around the room. Both Charles Wilson, the secretary of defense, and John Foster Dulles, the secretary of state, treated the question as a rhetorical one and didn't answer. There was no need for an answer. They all knew that the pressure would be great to consider such a flight the initial action in a nuclear attack and to launch an immediate response.

"What do we do if one of the planes is shot down and the pilot is captured?" Eisenhower continued. Richard Bissell, director of the U-2 program, began again to explain how fragile the airplane was and that it stood no chance of surviving a missile attack. Eisenhower knew that it was still a possibility. The U.S. had a great need for photographic intelligence, but this was not the way to acquire it. After spending several minutes tapping the eraser of a yellow pencil against a pad of paper, finally Eisenhower made his decision. The U-2 would never fly.

Without the U-2, Eisenhower was forced to increase the number of B-52 bombers on order. The United States had planned to build seven hundred of the flying monsters between 1954 and 1960, at a cost of $8 million each. Now, Eisenhower believed that the United States had to at least double that order. Without intelligence of the Soviet Union, there was no way to predict how many of the bombers might be successful if the United States were ever forced to launch an attack. There was no way to know where the Soviet antiaircraft missile installations were or where exactly their important military installations were located. By Eisenhower's guess, fewer than half of the B-52 bombers he might send over the Soviet Union would ever have an opportunity to drop their Cadillac-sized nuclear bombs.

That potential war with the Soviet Union seemed to be in its beginning stages just a year later.

In the summer of 1956, the president of Egypt, Gamal Abdel Nasser, had nationalized the Suez Canal, seizing it from the Suez Company, which was half owned by the British government. Eisenhower's advisors became concerned when in September the British and French had stopped sending intelligence reports to the United States, just as radio traffic between the two European nations increased. Secretary of State Dulles warned the administration, "I'm quite worried about what may be going on in the Near East. I don't think we have any clear picture about what the French and British are up to." On October 29, Eisenhower was awakened in the middle of the night. There were reports that Israeli paratroopers had dropped into the Sinai highlands overlooking the Suez Canal, and they had been supported by French fighter planes. The next day Eisenhower received word that the British had joined in the fighting against the Egyptians in the Suez region.

The British-French-Israeli attack on Egypt violated an agreement among Britain, France, and the United States not to interfere in affairs in the region. Perhaps of more concern, Nasser's government had an association with the Soviet Union, and there was the potential that the superpower might decide to intervene. An angry Eisenhower shouted at Dulles, "All right, Foster, you tell them that—God damn it!—we're going to do everything there is so that we can stop this thing."

On November 5, American intelligence agencies received urgent reports from several sources. British and French paratroopers had dropped into the Suez Canal, quickly followed by more soldiers in amphibious landing crafts. Israeli tanks were moving down from the Sinai to the Canal region. A full-scale invasion of Egypt was under way. "I just don't know what's gotten into these people," Eisenhower said. "It has to be the damnedest business I've ever seen: supposedly intelligent generals getting themselves into a position that they aren't going to be able to control at the end."

Later in the day, Eisenhower received an unusual offer from Moscow. The Soviets invited the United States to join them in stopping the British-French invasion of Egypt and claimed that they were considering dropping atomic bombs on London and Paris. Eisenhower knew

the Soviets were bluffing, and he also knew they were trying to drive a wedge between the United States and its
NATO allies.

The next day, November 6, 1956, was an election
day, and Eisenhower traveled to his home in Gettysburg,
Pennsylvania, so that he could cast his vote. The next
morning, the CIA informed Eisenhower that the Soviets
had reportedly told Nasser that they were sending fighter
planes to Syria to provide air support to the Egyptians.
This was the loss of control that Eisenhower had worried
about. The British and French should have known that
there was no way they could sponsor a war between
Egypt and Israel without the Soviet Union and the
United States becoming involved. If the Soviet planes
were in Syria, and if the fighters attacked British or
French troops, then by the requirements of NATO the
United States would have no choice but to attack the
Soviets. Within hours, the United States could be
involved in a global atomic war. With any luck, Eisenhower thought, he could contain the use of atomic
weapons to the Mediterranean region. "We have to be
prepared for every contingency," Eisenhower said. "Do
we at least know if the Soviets have ships in the Mediterranean? Somebody should check whether our ships there
have atomic antisubmarine weapons. This thing could
come from any direction."

There was no way to know if the Soviets were bluffing about the jet fighters in Syria, but Eisenhower had a
gut feeling that they weren't. "These boys are furious and
scared," he told Dulles. "Just like Hitler. It's just like
Hitler. We're going to force the Soviets into a position
where they believe their only option is to attack or else
they'll be humiliated.

"This is one dangerous situation. We better be damn sure that every intelligence point and every outpost of our armed forces is on their toes. If they start something, we'll have to hit them with everything in the *bucket*." Eisenhower sat back in his wingback chair and closed his eyes. In his imagination, he could see the silver Soviet MiG fighters on the hot desert tarmac, already fueled, with their engines sending up caloric waves in the heat as the pilots awaited the signal to begin the attack. If the Soviets attacked either the British or the French troops, that would be it. The United States was a member of NATO. If the Soviets attacked, then NATO would be at war. A war that the United States would lead, all because of outdated French and British colonial interests in Africa. This was not a war for Americans soldiers, Eisenhower thought. Not this time.

At a quarter after 6 P.M. that evening, just after the polls had closed in the Eastern time zone, the U.S. State Department issued a terse statement: The United States was withdrawing from NATO, effective immediately. With regard to nuclear war, the United States was NATO; without the United States as a member NATO was effectively dissolved. Each of the democratic nations would have to face the U.S.S.R. alone. That evening, just after Eisenhower told Dulles that he was going to go to bed, the secretary of state heard the president mutter under his breath, "God help us all."

At the Paris summit in May, Khrushchev cursed Eisenhower and then pointed to the sky and yelled, "I have been overflown!"

Charles De Gaulle then interrupted Khrushchev, saying that France, too, had been overflown.

"By your American allies?" Khrushchev asked.

"No, by you," DeGaulle replied. "Yesterday that satellite you launched just before you left Moscow to impress us over-flew the sky of France eighteen times without my permission."

Eventually, to save face, Khrushchev declared that satellite overflights were different from spy-plane flights and should be allowed. This unilateral change on the part of Khrushchev implemented the Open Skies policy that Eisenhower had pro-posed five years earlier. By August 1960, the first U.S. satellite was flying over the Soviet Union, and the days of U-2 flights over the forbidden territory were over. Francis Gary Powers was freed in February 1962 when he was exchanged on a bridge in Europe for Soviet spy Colonel Rudolph Abel, who had been caught spying in Brooklyn.

(In an odd and curious footnote to the U-2 episode, in October 1959, a Marine corporal who had worked on a U-2 base in Atsugi, Japan, defected in Moscow. Because Lee Har-vey Oswald had never been allowed into the U-2 hangar, and because he was treated so poorly by the Soviets, officials believed that he had not delivered any useful information about the U-2. "I don't think Oswald could have told them much they didn't already know," Bissell said years later. "Now what that has to do with the Kennedy assassination, God only knows.")

The United States and the Soviet Union almost did go to war over the British and French involvement in the Suez Canal conflict. On October 30, 1956, Francis Powers flying in a U-2 photographed puffs of smoke in the Sinai desert, which were evidence of the first shots fired in the conflict. The next day, another U-2 pilot flew over Egyptian airfields and pho-tographed the planes on the ground. He made a wide turn to return to his base and passed over the airfield a second time five minutes later. This time the airplanes and hangars were all burning—British forces had bombed the airfield while the U-2 pilot was making his turn. The United States sent the photographs via telephone to the British government to show that the U.S. knew that Britain was violating the Tripartite Agreement on the Middle East. With a full measure of pluck, the Royal Air Force sent back a thank-you: "Quickest bomb damage assessment we've ever had."

The Soviets were also having more fun than Eisenhower appreciated when they *did* invite the United States to join them in bombing London and Paris. Later, when the Soviets said they were sending jet fighters to Syria to confront the British and the French, the situation turned deadly serious. Eisenhower actually did inquire about the possibility of using atomic weapons against Soviet submarines in the area. He soon learned, however, that the Soviets were bluffing. U-2s had flown over the Syrian airfields and found no Soviet fighters. Instead of withdrawing from the NATO alliance or going to war, the United States joined with the United Nations to quickly bring a peaceful settlement to the conflict.

Two years later, the U-2 spy planes would again prevent a possible war. In September 1958, Chou En-lai's Communist Chinese government was threatening an invasion of Taiwan over two small, disputed islands, Quemoy and Matsu. The United States had pledged to protect Chiang Kai-shek's Taiwanese government, and it seemed that two of the world's largest nations were about to go to war over two small islands in the Pacific Ocean. But U-2 flights over China proved that the Communist Chinese had not moved any troops in place for an invasion, and so Eisenhower went on national television to reassure the American public: "There is not going to be any appeasement [of Chou En-lai], and there is not going to be any war."

In his memoirs, *Waging Peace*, Eisenhower pointed out that the negative information gained from U-2 flights was the most valuable the United States received. "Intelligence gained from this source provided proof that the horrors of the alleged 'bomber gap' and the later 'missile gap' were nothing more than imaginative creations of irresponsibility," he wrote. "U-2 information deprived Khrushchev of the most powerful weapon of the Communist conspiracy—international black-mail—usable only as long as the Soviets could exploit the igno-rance and resulting fears of the Free World."

Because Eisenhower was not drawn into an irrational and potentially very dangerous arms race, the United States was able to prosper. The most famous soldier of the twentieth cen-tury had best summed up what the enormous cost would have been to the nation if the United States had tried to keep up with the phantom military gaps. "Every gun that is fired, every warship launched, every rocket fired signifies, in the final sense, a *theft* from those who hunger and are not fed, those who are cold and not clothed," Eisenhower had said. "This is not a way of life at all, in any true sense. Under the cloud of threatening war, it is humanity hanging from a cross of iron."

The Failure of *Apollo 11*

———

Neil Armstrong and Buzz Aldrin risked everything to gain a Cold War advantage. What if they had failed to reach the moon?

LESS than three months into his young presidency, John Kennedy faced his first crisis on April 14, 1961. "What can we do?" he shouted about the latest chess move of the Soviets. "Is there any way we can catch them?"

In the 1960 presidential campaign, Richard Nixon had relied heavily on his image as a Communist fighter who could hold the Soviets in check. Now, just weeks into his term, Kennedy had been embarrassed by a one-orbit roller-coaster ride by Soviet cosmonaut Yuri Gagarin, the first man ever in space. The delicate perception of power in the Cold War between the United States and Soviet Union had shifted undeniably toward the Communists.

Kennedy remembered the hysteria that had followed the Soviet launch of the basketball-sized Sputnik satellite in 1957. Then, John Rinehart, a scientist at the Smithsonian Astrophysical Observatory, had said, "No matter what we do now, the Russians will beat us to the moon . . . I would not be surprised if the Russians reached the moon within a week." Another scientist, George Price, a former member of the Manhattan Project and a fellow of the prestigious American Association for the Advancement of Science, added, "Fewer predictions seem more certain that this: Russia is going to surpass us in mathematics and social sciences . . . In short, unless we depart utterly from our present behavior, it is reasonable to expect that by no later than 1975 the United States will be a member of the Union of Soviet Socialist Republics."

JFK wanted something done about the nation's space program, and he was willing to talk with anyone to see that it happened. "If somebody can just tell me how to catch up! Let's find somebody, anybody. I don't care if it's the janitor there, if he knows how." Kennedy was even willing to solicit ideas from his vice president, which for many presidents was going lower in the office hierarchy than the janitor. Kennedy asked Lyndon Johnson, "Do we have a chance of beating the Soviets by putting a laboratory in space, or by a trip around the moon, or by a rocket to land on the moon, or by a rocket to go to the moon and back with a man? Is there any other space program . . . in which we would win?"

Kennedy found his goal and, in a speech to Congress just weeks later, declared that the Americans would be the first to place a man on the moon and that they would do it by the end of the decade. When he announced this, Kennedy told Congress that the mission would influence "the minds of men everywhere who are attempting to make a determination of which road they should take." The idea that NASA could make enough technological advances to send a man to the moon seemed almost laughable—at the time, NASA transported Mercury capsules to the launching pad on the back of a flatbed truck on a pile of mattresses. In fact, NASA's own study of the mission shortly after Kennedy's announcement predicted that the United States stood only a one-in-ten chance of making a successful landing by Kennedy's deadline.

❧

In June 1969, _Apollo 11_ astronauts Neil Armstrong and Buzz Aldrin were in a lunar module at Houston's mission control, heading toward the projected image of the moon, when suddenly a thruster stuck on. The lunar module began spinning like a county fair tilt-a-whirl ride. Aldrin hesitated as Armstrong struggled to regain control, but finally yelled, "Neil, hit abort!" Mission control agreed, saying, "_Apollo 11_, we recommend you abort." But before Armstrong could respond, the lunar module shattered into a million pieces as it crashed into the moon. Aldrin looked over at Armstrong, who was still trying to think his way out of the problem.

It had been just one of many unsuccessful simulations, but the simulation had been real enough, and near enough to the date of the actual flight, that it caused tension between the two men that wasn't completely healed by the July launch.

Armstrong had been known to be one of the best pilots at NASA. In 1966, Armstrong was at the helm of _Gemini 8_. The mission of that flight was to dock two spacecraft together—one of the many incremental steps that led to the moon landing. But as soon as the two spacecraft joined together, they began spinning. Armstrong separated the two space vehicles, and then trouble really began. _Gemini 8_ began spinning at more than one revolution per second, and it was in danger of spinning off into space. Inside the craft Armstrong and David Scott found that their hand controllers were useless. The two astronauts could not focus their vision, and Armstrong deadpanned, "I gotta cage my eyeballs." As the g-forces increased, the two men were about to black out. Finally, Armstrong activated the reentry control system and, after a moment, regained control of his hand controller and was able to stabilize the craft.

Saving *Gemini 8* had required great physical ability, quick thinking, and expert piloting. It was hardly a surprise when NASA picked Armstrong to pilot the first real moon mission.

Just before the mission began, NASA administrator Thomas Paine met with Armstrong and the other two men on *Apollo 11*, Buzz Aldrin and Michael Collins, and made a surprising offer: If they needed to abort the mission, they could do so without losing their place in line for a chance to go to the moon. It was a break with NASA tradition and policy, but Paine wanted to emphasize that the crew wasn't to take any unnecessary chances to be the first to make the moon.

The men who would actually land on the moon, Aldrin and Armstrong, were quietly confident, but the more outspoken Michael Collins told NASA officials, "I am far from certain that we will be able to fly the mission as planned. I think we will escape with our skins, or at least I will escape with mine, but I wouldn't give better than even odds on a successful landing and return. There are just too many things that can go wrong." Armstrong reportedly agreed that the chances of success for *Apollo 11* were no more than fifty-fifty.

⌘

The Soviets fully understood the difficulty of landing a man on the moon, and although they were behind in the race to the moon, they were poised to pass the Americans if NASA stumbled. Officially the Soviet Union denied that it had a lunar landing program, but the denials were considered nonsense by NASA personnel, who had learned several details about the

Soviet plan. The Soviet method had its differences from that of the Americans—instead of launching two vehicles together, for example, the Soviet launched the lunar lander and the command module separately, and then the cosmonauts transferred to their lunar module by performing a spacewalk. Another difference was in the reentry: The Russian command module was slowed by skipping across the top of the atmosphere like a rock skipping across the surface of a lake, instead of dropping straight through as American spacecraft did. But in their broader aspects, the plans to reach the moon were very similar.

On the fourth of July, 1969, less than two weeks before the scheduled launch of *Apollo 11*, an American Keyhole spy satellite photographed two rockets being prepared for flight at the launch facility in Tyuratam, Kazakhstan. One was a G1 rocket topped with a Soyuz lunar lander. Nearby was a Proton rocket, which would lift a second Soyuz spacecraft and three cosmonauts a day later. The next morning the Keyhole spy satellite photographed the G1 rocket being fueled, and CIA and NASA analysts expected the rocket to be launched at midday. Less than two weeks before the launch of *Apollo 11*, the Soviets were going to make a moon launch. Officials at the CIA and NASA wondered, was this a prelude mission, like *Apollo 9* or *Apollo 10*, or was this a desperate attempt to beat the Americans to the moon?

In the end, the purpose of the Soviet mission didn't matter. That afternoon, when the Keyhole satellite passed over again, it photographed a smoldering scene of destruction, a black kidney-shaped scar on the Kazakhstan steppe that was several miles in diameter.

After the G-1 had been fueled, an electrical short in the third stage at the top of the rocket had ignited the fuel. As Soviet controllers and cosmonauts watched from the safety of bunkers several miles away, the rocket had exploded into an enormous fireball, destroying not only the rocket but also the launch facility. The explosion would set back the Soviet manned moon program at least a year.

◈

On July 20, 1969, after a four-day flight, the American lunar module, *Eagle*, approached the moon. The position of the ungainly craft was confirmed by what the Midwesterner Armstrong referred to as "barnyard math" and the assistance of a low-tech stopwatch. As the *Eagle* neared the surface, Armstrong's calculations showed that something was wrong. The *Eagle* wasn't going to land where it was supposed to. "Okay, we went by the three-minute point early. We're long," Armstrong reported to Houston control. "Our position checks down range show us to be a little long." Actually, they had missed the target-landing site by three miles.

Before Armstrong and Aldrin could dwell too much on their position, a critical problem appeared. A warning light on the craft signaled that something was not working right. "Program alarm," Armstrong reported, and Aldrin saw his data screen go blank.

"It's looking good to us. Over," Houston control replied, but Armstrong repeated, "It's a *1202*."

The problem was that neither the astronauts nor Houston control was immediately sure what a 1202 alarm was. Houston quickly determined that the *Eagle's* computer was reporting that it was overloaded.

The *Eagle's* computer was, of course, primitive by modern standards. It held 70 kilobytes of ROM and 4.4 kilobytes of RAM; a typical home computer of 200 megabytes in the year 2000 would be about fifty thousand times more powerful. The lunar lander's computer was set up so that it ran through a list of tasks, and if it was not able to complete a task it would reboot and begin the list again. If it was unable to get past this roadblock, it would activate the Alarm/Abort routine. This scenario had been practiced in the simulator, with unfortunate results. On July 5, just eleven days before the launch of *Apollo 11*, the mission control engineers had been running through a simulation with the crew of *Apollo 12* in the simulator, as the crew of *Apollo 11* looked on. During the simulation, an alarm sounded that had never appeared before, a computer alarm, and the engineers and crew were mystified about how to respond. In the simulation, with the lunar module ten thousand feet above the lunar surface, the engineers had decided to abort the mission.

After the simulation NASA had instructed the controllers to learn each of the computer alarms and to figure out a way to respond to them. The problem was that the computer alarms were true "idiot lights": They simply indicated that a problem existed—there was no way to tell if the problem was insignificant or catastrophic. However, the engineers considered the worry over the alarms just institutional paranoia. "The problems that triggered the alarms were not problems that reasonably could happen during a mission," one engineer said. Nevertheless, a crude protocol was worked out. Because the cause of the alarm couldn't easily be determined, the engineers decided that if an alarm sounded just once, they would ignore it, but if it stayed on or repeated, they would abort the lunar landing.

As the alarm light continued, both astronauts focused their eyes on the large red Abort Stage button.

"What is it? Let's incorporate . . ." Armstrong paused, not knowing what to do next. Then he put the question to mission control: "Give us a reading on the 1202 program alarm."

"Give us a reading" was an astronaut's way of asking,

"Should we abort?" The crew was growing understandably impatient waiting for a response.

To abort the mission, one of the astronauts simply had to hit the Abort Stage button, which would fire pyrotechnic bolts holding the two sections of the craft together and fire the ascent engine. If one of the bolts didn't fire, or if the rocket engine didn't light, the spacecraft would crash.

After eleven seconds of indecision, Houston gave Armstrong and Aldrin a stuttering, not altogether reassuring reply. "We've got you . . . We're go on that alarm."

Within two minutes, Aldrin had figured out an override that would prevent the computer from overloading, but just a minute later, another alarm began sounding, this one a *1201*. The engineers at Houston knew that this alarm was similar to the *1202* alarm, and they quickly responded (this time with more confidence), "We're go. Same type. We're go." Aldrin was not completely reassured, and his muscles were rigid from tension.

The computer problem had apparently been solved, but at a dear price. Because of the danger of an impending abort, Armstrong and Aldrin had focused their attention on correcting the alarm instead of evaluating possible landing sites. In the simulations Armstrong had checked for landmarks along what the astronauts called US 1, the path to the Sea of Tranquillity that was named after the well-known highway that runs along the Atlantic coast. When the *Eagle* was just two thousand feet above the moon's surface, Armstrong finally began looking for a place to set the ungainly craft down and saw that he was past all of the landmarks he had carefully memorized from photographs.

The lunar module's automatic guidance system was taking the *Eagle* down toward the edge of a large crater, in a field of car-sized boulders. For a moment, when the craft was just five hundred feet above the surface, Armstrong considered squeezing the lunar module between the boulders because he thought that the site obviously offered geologic variability that would be interesting to the scientists back on Earth. He quickly decided that it wasn't worth the risk. "Pretty rocky area," Armstrong reported with the test pilot's traditional economy of words. "I'm going to . . ." Armstrong took control of the craft from the automated system and pitched the *Eagle* over to fly it over the crater. Back in Houston, mission control could tell from its instrumentation that Armstrong had taken manual control of the craft even though he hadn't reported it. Realizing that the crew had their hands full, mission control was ordered to be silent.

Forty-five seconds after taking manual control of the craft, Armstrong asked, "two hundred seventy [feet]. Okay, how's the fuel?"

"Take it down," Aldrin replied. Sixty seconds after the switch to manual control, another warning light sounded, this one caused by the landing radar, which couldn't get a fix on the lunar surface because the *Eagle* was tipped over in forward flight. It was promptly ignored. But as the vehicle approached one hundred feet—ninety seconds after Armstrong bypassed the computer-selected landing site—a more serious alarm sounded. "One hundred feet, three and a half down, nine forward. Five percent [of fuel left]. Quantity light," Aldrin reported.

The amber Descent QTY was a fuel warning light, indicating that they had sixty seconds of fuel left in the tank, of which twenty seconds were supposed to be reserved for a possible abort. Nervous ground controllers at Houston broke their silence to report what the astronauts already knew: "Sixty seconds."

"Light's on," Aldrin reminded Armstrong, who was still looking for a place to land. Thirty seconds after the quantity light first came on, Aldrin noted, "Picking up some dust." As the craft descended to forty feet above the lunar surface, it began blowing up the talcum-powder–like surface of the moon. Armstrong found flying in the dust "a little bit confusing." He was flying blind and running out of fuel. As he eased the craft toward the surface, it began to unexpectedly shift around, first to the right, then backward and to the left. "You don't like to be going backwards, unable to see where you're going," Armstrong later said in his NASA debriefing. "I was also reluctant to slow down my descent rate any more that it was, or stop [the descent], because we were close to running out of fuel. We were hitting our abort limit."

Armstrong had two unpleasant choices. He could abort the mission because of the danger of running out of fuel, or he could run out of fuel and let the machine fall the last twenty feet or so to the surface, which could damage the craft and make a return to Earth impossible. His decision was made. Armstrong kept flying.

"Thirty seconds," mission control reminded, and Aldrin again looked at the Abort Stage button. Ten, fifteen, twenty seconds passed by and Armstrong was still flying.

Then, with the computer showing just five seconds of fuel left, a blue light on the instrument panel began to glow, and Aldrin announced "Contact light."

For just a moment Armstrong didn't realize what Aldrin had said, and the craft shifted to the left as Armstrong continued to fly. Then, three seconds later, he shut down the engines. Even then the astronauts didn't relax: They immediately began preparing the spacecraft for takeoff. The concern was that if something had been damaged in the landing, and the astronauts needed to abort the mission, they should do that immediately. There was also the small concern that the craft might be unstable on the surface—one prominent U.S. scientist had insisted that the surface of the moon was a thin crust that would collapse under the weight of the lunar craft—and this might also force an immediate evacuation. Seventeen seconds after Aldrin announced that they had reached the surface, however, the controllers at mission control couldn't contain their silence any longer. "We copy you down, *Eagle*," they prompted.

Armstrong paused as he continued the shutdown procedures, and then replied, "Houston, Tranquillity Base here. The *Eagle* has landed."

WHAT IF?

What if the crew of Apollo 11 *had decided to abort their mission?*

Armstrong was guiding the lunar landing module toward the surface of the moon when, suddenly, flying dust blocked his view. The *Eagle* lander didn't have large windows like an airplane—it had only one small porthole on the side. To land the craft, Armstrong needed to look out that window and choose a surface landmark in the distance, and ease the craft down, using that landmark as a reference point to determine his angle and altitude. He instinctively increased the thrust to lift the craft up off the surface so that he could regain his reference point, which was just one boulder in an ocean of rocks.

"Fifteen seconds," the voice from mission control said, reminding Armstrong of the fuel situation. Armstrong continued staring out the window at his preferred reference boulder. Even though he couldn't see him, he could somehow sense Aldrin's tension. Slowly, carefully he began backing the *Eagle* toward the surface of the moon as the dust began rushing past his small window again. He was taking the craft down slowly, deliberately, so that he could maintain a level descent in the blinding dust. A hard landing—or even worse, tipping the *Eagle* over on its side—would guarantee that the lifeless surface of the moon would be their final stop. Armstrong felt as if he were walking through a completely dark room with his arm outstretched, hoping to find a wall or something to hold onto, hoping that soon he would feel the thud that would tell him that the *Eagle* had found the surface. Instead, as the seconds went by, all he could feel was the craft slightly rocking on the point of its thrust. Armstrong could tell that time was getting short, maybe ten seconds, nine seconds, and he was still stretching his hand out, hoping to touch something, eight seconds, seven seconds, still flying, six seconds, five seconds, gently rocking, four seconds, reaching out, three seconds.

Armstrong's gloved hand shot out toward the red abort button, and instinctively Buzz Aldrin's hand jumped forward and landed on top of Armstrong's. Their hands still together, the two astronauts braced themselves against the console as the explosive bolts fired and the lunar lander reversed itself and shot into the lunar sky. The Ohio farmboy Armstrong later said in his debriefing that the separation felt like the first time he had fired a twelve-gauge shotgun when he was seven years old.

Just as Armstrong and Aldrin felt relief that the lunar lander hadn't continued downward to crash onto the lunar surface, the lunar lander suddenly began swinging wildly from side to side. "Son of a bitch!" Armstrong yelled, violating NASA's rule against profanity on the radio. Afraid that the craft was about to cartwheel out of control and crash onto the moon, Armstrong assumed manual control and was able to keep the lander reasonably stable. The craft would begin swinging again each time the astronauts tried to let the computers assume control, which they needed to do in order to locate the orbiting command module in the vastness of space. Finally Aldrin discovered that the radar receiver had inadvertently been turned on in the excitement over the abort. The switch wasn't to be activated until the craft was within radar range of the command module. With no radar signal to lock on to, the *Eagle*'s computer had directed the craft into a crazy series of sweeps across the lunar sky. With the switch off, the craft was able to follow the computer's commands.

It wasn't surprising that the wrong switch had been activated during the abort. Before the mission Armstrong had stubbornly refused to practice aborts in the NASA simulator. Like Cortés burning his boats on the Mexican shore as he began his conquest, Armstrong didn't want to give himself a way out when the situation appeared impossible. Because the crew hadn't drilled the abort procedure, however, they nearly lost their lives.

When the *Eagle* docked with the command module, Aldrin was the first through the tunnel and into the cabin. Michael Collins, who had remained with the command module in orbit above the moon, didn't know what to say to Aldrin and simply shook his hand firmly. Armstrong immediately followed through the tunnel. The three men had work to do to get the command module back to Earth, welcome work, work that kept them from discussing what had happened on the lunar surface below.

When the crew splashed down in the Pacific Ocean, they were greeted by an odd, subdued celebration. No one wanted to bring any additional embarrassment to the *Apollo 11* crew, and so the naval band still played and the men still cheered when the crew stepped onto the deck of

U.S.S. *Hornet*. Even President Richard Nixon, who had traveled to the Pacific to join in the *Apollo 11* celebrations, continued with the plans and greeted the *Apollo 11* crew. The president seemed upbeat and, laughing, told the three, "I'm a Redskins fan, and so I know what it's like to say, 'We'll get 'em next time!' And I'm sure you'll be back up there soon."

Luckily for NASA, there would be no need to make Nixon out to be a liar: They had already made the decision to send the *Apollo 11* crew up again in the next lunar landing mission. There was one change, however: Buzz Aldrin would now be the crew commander.

Neil Armstrong was NASA's best pilot, but there were some at NASA who believed that Aldrin was NASA's best all-around astronaut. NASA officials were careful not to suggest that the aborted mission had caused a demotion of Armstrong. "Look," NASA administrator Thomas Paine said, "if we had any concerns about his ability, we wouldn't be sending him back up there. We're not going to get too many chances to do this."

NASA officials wanted to send another practice mission to the lunar surface before sending Aldrin and his crew back to the moon—a repeat of *Apollo 10*. But Nixon and his aides were apoplectic at the thought of having the historic lunar landing mission be named *Apollo 13*. The White House had Vice President Spiro Agnew call Paine to express President Nixon's concerns. NASA officials decided that another run-up mission wasn't necessary anyway, and the *Apollo 12* mission was scheduled for October.

Early on the morning of October 4, 1969, the *Saturn V* rocket thundered and threw Aldrin, Armstrong, and Collins back toward the moon. This time, the lunar lander—nicknamed "Liberty Bell" as a tribute to astronaut Gus Grissom's first space flight in *Gemini* 7—had a more accurate fuel system, and the crew was more confident. NASA and the crew had decided to go back to the landing site that Armstrong had selected so that as many landmarks as possible would be recognizable to the astronauts. *Apollo 12*'s Liberty Bell successfully landed on the moon's surface on October 6. The mission schedule had called for Aldrin and Armstrong to sleep immediately after landing on the moon so that they would be well rested for the moon walk the next morning, but the two men convinced the flight directors at mission control that they were too excited to go to sleep. Just after 11 P.M. Eastern Daylight Time, the two astronauts were ready to step out of Liberty Bell and onto the moon's surface.

"We come not just as Americans, but as men from Earth," Aldrin said as he jumped the last meter to the lunar surface and became the first human to walk on the moon. "We come in peace for all mankind."

Although the failed mission of *Apollo 11* had gone well for NASA (with the possible exception of Neil Armstrong, who was denied his chance to be the first on the moon), it caused a dramatic shift in the foreign policy of the Nixon administration. Nixon's earlier trip to the Pacific (code name: *Moonglow*) wasn't made simply to congratulate the *Apollo 11* astronauts—the trip had a secret mission, too. Henry Kissinger, the nation's national security advisor, had accompanied Nixon. After the greeting of the astronauts, while the world's attention was focused on their success, Kissinger was to peel off and travel to Paris for secret peace talks with the North Vietnamese.

John Ehrlichman had said that Nixon viewed the Apollo program in the same category as ceremonies for

returning servicemen and the Prussian-military–style uniforms that Nixon had briefly dressed the White House
police in. In other words, just more national bunting for a
Fourth of July celebration. Which wasn't to say that
Nixon thought that the Apollo program was unimpor-
tant—he realized the value of ceremony. It was just something that he didn't care to waste any of his time on. And
in the months before the successful *Apollo 12* mission, the
failure of *Apollo 11* eclipsed everything else he looked at.

Just two months before, he had announced a peace
plan for Vietnam. But now Nixon knew that he couldn't
work for peace because the reputation of the nation was at
stake.

"Now we can't do it out, Henry," Nixon turned and
said to Kissinger. "We talk with the North Vietnamese
now, the Soviets and the Chinese will think we're chicken
****s if we try to pull out of this thing. We can't get to the
moon, and we can't hold back the North Vietnamese.
We're going to be laughingstocks around the world."

"Not laughingstocks, Mr. President," Kissinger said.
"No one would laugh at your visionary leadership and the
way . . ."

"That's right, they're not going to laugh at us, Henry.
Not at Nixon. We're not going to Paris. We're going to
get our act together at NASA, and we're going to get our
act together in Vietnam, and we're going to win them
both."

Instead of beginning peace negotiations, Nixon
ordered more bombings of North Vietnam. Within weeks,
when it became obvious that the North Vietnamese were
being supplied by the Chinese through the neighboring
countries of Cambodia and Laos, Nixon ordered massive
bombings of those countries, too. In a nationally televised

address, Nixon defended his expansion of the war into two additional countries. "Whether I may be a one-term president is insignificant compared to our failure to act in this crisis," Nixon said. "I would rather be a one-term president and do what I believe is right than to be a two-term president at the cost of seeing America become a second-rate power and to see this nation accept the first defeat in its proud 190-year history."

Nixon's fears about his political future were correct, much to his disappointment. The expansion of the war, together with the reversal of troop withdrawals, caused Nixon's support to wither, even among some Republicans.

When Nixon had taken office, several people, including Democratic New York senator Patrick Moynihan, had urged him to pull out of Vietnam immediately. The thinking was that if Nixon admitted early in his term that the war in Vietnam was a failure and withdrew American troops, the loss would be blamed on Lyndon Johnson, who had started the ill-fated war. But Nixon refused to pull out immediately, preferring instead a slow gradual withdrawal that the Nixon administration called "Vietnamization." When he abandoned this plan after the failed flight of *Apollo 11* and began ratcheting up the United States' involvement in Vietnam, the war became Richard Nixon's war, and that gave his opponents a target to shoot at. As Nixon continued to insist during the 1972 election that the credibility of the United States was at stake in Vietnam, enough voters decided that Mr. Nixon's war wasn't worth fighting that his Democratic opponent, George McGovern, was elected to replace him as president.

In 1906, President Teddy Roosevelt said that he was going to send the U.S. Navy on a forty-six-thousand-mile cruise around the world. Britain and German naval officials, who commanded two of the world's preeminent navies, said that it was impossible for such a large fleet to sail that far. Roosevelt ordered the worldwide cruise anyway. (In a typical TR touch, when Congress balked at paying for the world navy tour, Teddy had told them that he had enough money to send the ships halfway around the world and that if Congress wanted their navy back, they should appropriate the rest of the money.) The cruise was a rousing success, and it established the United States as a military superpower. At the end of his administration, Roosevelt had said, "In my own judgment, the most important service that I rendered to peace was the voyage of the battle fleet around the world."

Brandishing the big stick of American technological superiority has been an important part of maintaining the United States' position as a superpower through the twentieth century.

After President Nixon greeted the successful *Apollo 11* crew, National Security Advisor Kissinger did continue on to Paris, where, flush with the pride of *Apollo 11*, he began the first negotiations to end the Vietnam War. The negotiations didn't go as well as planned (Kissinger once described his North Vietnamese counterparts as "tawdry, filthy s***s"), but the talks did bring about a gradual pullout and, on January 27, 1973, a cease-fire in the Vietnam War.

Although *Apollo 11*'s lunar lander, the *Eagle*, had appeared nearly out of fuel as it approached the moon's surface, NASA discovered later that the *Eagle* had had more fuel than was thought. Its computer had showed that there was only five seconds of fuel left, when in fact there was fifty seconds of fuel left—still not much, but a world of difference to a skilled pilot like Armstrong. The misreading was the result of a mechanical problem instead of a computer error: When Armstrong pitched the *Eagle* over to fly across the crater that was in the landing area, the fuel had sloshed around. In the reduced gravity of the moon, the fuel didn't settle quickly, and the sensors had misread the amount of fuel left. In all the *Apollo* missions that followed, the fuel tanks contained antisloshing baffles.

Armstrong was determined to land the *Eagle* on the moon. No doubt he viewed it simply as accomplishing his mission, but if *Apollo 11* had not been successful, the United States might have been drawn further into an unfortunate war.

Chapter 25

The Impeachment Trial of Richard Nixon

❧

President Richard Nixon decided to resign rather than face an almost certain impeachment and conviction by Congress. What if he had decided to fight instead?

IN the *U.S. v. Richard Nixon*, Nixon lost. It was a unanimous decision.

On July 24, 1974, the U.S. Supreme Court issued a unanimous knockout that instructed President Nixon to turn over audio recordings that had been made in the Oval Office. Up to that point, Nixon had denied knowing about the numerous illegal activities that had gained the general moniker "Watergate." But the tapes proved Nixon's involvement, and so Nixon lost not only his case in the Supreme Court, but his presidency as well.

"A Nixon-Agnew administration will abolish the credibility gap and reestablish the truth, the whole truth, as its policy," Vice President–elect Spiro Agnew had said in 1968, as he and Nixon anticipated taking their positions as the nation's executive officers. Five years later, Nixon would become famous for telling his staff, "I don't give a s*** what happens. I want you all to stonewall it."

～～～

"Operation Gemstone" was the name of the attempted bugging of the Democratic National Committee (DNC) headquarters by G. Gordon Liddy and E. Howard Hunt, who worked for the Committee to Reelect the President (an organization that would gain the unfortunate acronym CREEP). The two former intelligence agents—Liddy had worked for the FBI and Hunt for the CIA—had sent five men into the Watergate Hotel in Washington, D.C., to break in to the DNC offices to place microphones. Just after midnight on June 17, 1972, security guard Frank Wills noticed a piece of tape across the lock on a door that prevented the door from locking. Wills removed the tape, but when he passed by again and saw that the tape had been replaced, he called the police. Washington police soon arrived and arrested five men for the break-in.

"That is not the beginning," presidential attorney John Dean would later explain to Sam Dash, the chief counsel of the Senate committee investigating the break-in. "You are making the mistake of concentrating on the break-in of the Democratic National Committee headquarters. Frankly, that was not very significant when viewed in its total context." Over the next two years, Americans would discover that the "third-rate alleged burglary attempt," as presidential spokesman Ronald Ziegler dismissed the scandal, was just one example of a widespread use of illegal activities aimed at keeping Richard Nixon in office.

From the time the Watergate burglars were arrested, the White House denied that any of the men working for CREEP had any connection with the White House or that anyone in the White House had any knowledge of their activities, and most Americans believed the story. "As shameful as Watergate is," the *Orlando Sentinel* said in a January 1973 editorial, "it has a hopeful or reassuring aspect: nothing is being swept under the rug."

When the seven men involved were being sentenced in March 1973, however, two of them, James McCord Jr. and Jeb Stuart Magruder, told the judge that they had been working for the White House. Three weeks later, after damaging reports in The *Washington Post* implicated White House staffers, Ziegler was forced to issue a new White House version of events. "This is the operative statement. The others are inoperative," was how Ziegler told reporters that the White House had been lying.

Two weeks later Nixon said that he took full responsibility for the improper actions of his staff, even though he had never had any knowledge of their activities. That same day Harvard law professor Archibald Cox was appointed special prosecutor to look into Watergate. Cox's prosecutor office wasn't the only group looking into Nixon's involvement in the illegal activities. The Senate Select Committee on Presidential Campaign Activities, chaired by North Carolina senator Sam Ervin, was also holding televised hearings on Watergate. By midsummer, Republican minority leader Howard Baker of Tennessee concisely asked what became the crux of Watergate: "What did the president know and when did he know it?"

But Watergate was about more than just the break-in at the Watergate Hotel. There were other break-ins, such as the one of the office of the psychiatrist treating antiwar activist Daniel Ellsberg; illegal wiretaps; ignored subpoenas; suspicious transfers of money, including hush-money payments;

campaign dirty tricks, such as letters offered to reporters as "evidence" that Henry "Scoop" Jackson was a homosexual and that Hubert Humphrey was patronizing call girls; probable tax evasion by the president (for two years of his first term as president, Nixon had paid less than one thousand dollars in income taxes); and illegal campaign contributions.

And it had all been caught on tape.

On July 16, 1973, a White House staffer named Alexander Butterfield revealed that all conversations in the president's office had been secretly recorded. The fuse on the Nixon presidency had been lit.

A week later, both Cox and the Senate committee subpoenaed the tapes. Nixon refused to hand them over, citing executive privilege and national security concerns. Cox considered using a group of U.S. Marshals to go into the White House to seize the tapes, but he changed his mind, worried about what might happen if the marshals were opposed by the White House marine guards.

When two higher courts upheld Cox's subpoena, Nixon offered to provide written summaries of the tapes. Cox refused to accept the substitution. On Saturday, October 20, 1973, Nixon ordered his attorney general, Elliot Richardson, to fire Cox. Richardson resigned instead. Nixon then ordered the deputy attorney general, William Ruckelshaus, to fire Cox, but Ruckelshaus resigned. Finally, solicitor general Robert Bork, newly in charge of the Justice Department, agreed to fire Cox. The Saturday Night Massacre, as the event came to be known, severely hurt Nixon's political standing both among the voters and among members of Congress.

As the weeks passed, however, Nixon remained defiant about his lack of involvement in the illegal activities. At a press conference at Disney World in Orlando, Florida, in November, Nixon gave a Mickey Mouse denial: "I welcome this kind of examination because people have got to know whether or not their president is a crook. Well, I'm not a crook."

In May 1974, as Nixon's argument for withholding the tapes because of executive privilege made its way through the courts, U.S. district court judge John Sirica ordered Nixon to turn over all remaining tapes to Cox's replacement, Leon Jaworski. Nixon then took his case to the Supreme Court, and on July 24, the justices voted eight to zero that the president had to turn over the evidence.

At this point, Nixon faced the most difficult decision of his eventful political career.

He had listened to the tapes. On one tape that contained a conversation between Nixon and Haldeman on June 23, 1973, Nixon could be heard telling Haldeman that they needed to get the Justice Department and the FBI to back off the investigation. Haldeman had suggested that he and Ehrlichman call CIA director William Helms into the White House and urge him to have the CIA tell the FBI that the Watergate break-in and related activities were all part of a CIA operation, and that the FBI should end its investigation. "All right, fine," Nixon said. Later on the tape, Nixon tells Haldeman how to handle the CIA officials: "Play it tough. That's the way they play it and that's the way we're going to play it."

"Okay," Haldeman replied. "We'll do it."

"Say, 'Look, the problem is that this will open the whole, the whole Bay of Pigs thing, and that the president just feels that . . . Ah, without going into the details," Nixon instructed Haldeman. "Don't . . . don't lie to them to the extent to say there is no involvement. Just say, 'This is a comedy of errors,' without getting into it. 'The president believes that this is going to open the whole Bay of Pigs thing up again.'" And, ah . . . 'Because these people are plugging for keeps,' and that they should call the FBI in and say that we wish for the country: 'Don't go any further into this case. Period!' "

Nixon had not only approved of the cover-up, he had provided the script. The June 23 tape was the "smoking pistol" that the investigators had been looking for over the previous two years. Once the tape was turned over to the special prosecutor, there would be nothing left to deny. Richard Nixon would become the first president to be impeached and removed from office.

Now that the Supreme Court had ordered him to turn over the tapes, he had at least three choices. He could comply with the order and face certain impeachment and removal from office. He could refuse to turn over the tapes and face *almost* certain impeachment and likely removal from office. Or he could resign and save the nation the turmoil of an impeachment trial. Several of his aides began moving him toward a position where he could easily do the latter.

But Nixon's family was opposed to his resignation, and he himself swayed from one extreme to another. On Friday, August 2, 1974, Nixon told Alexander Haig, his chief of staff, that he had changed his mind about resigning. He had listened to the tape again, he told Haig, and he decided that his comments to Haldeman on June 23 weren't as bad as they first sounded. "Let them impeach me," Nixon told Haig. "We'll fight it out to the end."

Nixon was inclined to remain in office on principle. On Tuesday, August 6, Nixon made a statement in a meeting with his cabinet: "I am of the view that I should not take the step that changes the Constitution and establishes a precedent for all future presidents. It would result in a parliamentary system with all its weaknesses but none of its strengths."

However, Nixon assured his cabinet, it basically came back to the idea that what he had done wasn't an impeachable offense. "If I thought there was an impeachable offense, I wouldn't put the Senate through the agony of trying to prove it."

Then Vice President Gerald Ford interrupted the president and asked to speak. This was a violation of the unwritten rules of cabinet meetings, but a surprised Nixon conceded the floor. Looking directly at Nixon, Ford told him, "Had I known what has been disclosed in reference to Watergate in the last twenty-four hours, I would not have made a number of the statements I made either as minority leader or as vice president." Then Ford began speaking as if Nixon's resignation were a fait accompli. "Let me assure you I expect to continue to support the administration's foreign policy and the fight against inflation."

Nixon sat with a frozen half-smile on his face, ignored Ford's statement, and began a discussion of the economy, but everyone in the room was aware of a commotion along the wall of the room. George Bush, the chairman of the Republican National Committee, who was attending the cabinet meeting as Nixon's guest, was trying to be recognized to speak. Nixon pointedly ignored him, but Bush interrupted the president and spoke anyway. The country was suffering and public support was vanishing, Bush said; the time had come for Nixon to resign. A shocked silence fell over the room, and when Nixon again tried to discuss the economy, William Saxbe, the attorney general, said that before anything was done on the economy, "We ought to be sure that you have the ability to govern."

Nixon had called the largest possible cabinet meeting, hoping that in such a large group people would be afraid to appear disloyal. Once one or two expressed their support for him, he thought he could get everyone in the room to join in.

Instead, three of the most important attendees had all but insisted that he resign. For a moment it appeared that this would be the consensus of the people in the room. The cabinet meeting coup was derailed only when Secretary of State Henry Kissinger stood and said in a raised voice, "We are here to do the nation's business!" However, after the meeting, Kissinger told Nixon in private that he also thought that he should resign.

A little later, Nixon told Haig that he was still defiant. "Obviously I can resign and save everybody a lot of time and trouble. That's a temptation. On the other hand, Al, I may just run it out, go through the Senate trial, put my defense on the record, accept conviction if that's what's in the cards, go on trial, go to prison, lose everything, but go out with my head held high." Later that evening, Nixon was morose. "You know, Al, you soldiers have the best way of dealing with a situation like this," the president told his chief of staff. "You just leave a man alone in a room with a loaded pistol." David Eisenhower, Julie Nixon's husband, thought it was possible that the president might commit suicide, saying that he had been "waiting for Mr. Nixon to go bananas."

Despite her husband's concerns, Julie Nixon Eisenhower was vocal in her opposition to resignation, even if it meant removal by the Senate. "Daddy's not a quitter," she told speechwriter Pat Buchanan. "It would be better for Daddy if he laid it all out before the country. . . . He could stress his accomplishments against this smaller, less significant thing."

The next day, Nixon's outlook improved only slightly. "Lenin and Gandhi did some of their best writing in jail," he told one of his aides. "I've never quit before in my life. That's what nobody around here has understood during this whole business. You don't quit!"

On Saturday, July 27, the House Judiciary Committee met to vote on articles of impeachment. President Nixon was at his home in San Clemente, California, while his staff monitored the House proceedings. Just after seven o'clock that evening, the House committee passed the first article of impeachment. White House press aide Diane Sawyer ran to tell her boss, press secretary Ziegler, so that he could inform the president. "The poor president," Sawyer wailed. "Oh, the poor man."

Eventually the committee would pass three articles of impeachment against the president. Two additional impeachment articles—one pertaining to the secret bombing of Cambodia and one accusing Nixon of tax evasion—were rejected. Next the three articles of impeachment would be sent to the full House for consideration.

Back at the White House, chief of staff Al Haig began hardening the troops for battle. Haig asked speechwriter David Gergen if there were any "weak sisters" on the speechwriting staff who might leak information to the press. If so, they should be fired, Haig said. "I'm probably the weakest of all," Gergen said.

Some Republicans were holding firm: James Eastland, senator from Mississippi, had promised Nixon in the midst of the scandal, "I don't care if you're guilty or innocent, I'll vote for you." But among most Republicans, support for the president was evaporating. Bob Dole and John Tower, two Senate leaders, were whispering that they might vote against Nixon if the House voted to impeach the president.

With little support in Congress or in the general public for him to stay in office, Nixon finally decided to resign. On August 8, 1974, the evening his resignation was to take effect, Nixon surrounded himself with a pile of books. They were the autobiographies of other presidents, and Nixon spent the evening seeking solace and, possibly, forgiveness from his peers.

WHAT IF?

What if Nixon had refused to resign following the House Judiciary Committee vote to impeach him?

On Friday evening, August 9, 1974, just after the nightly news broadcasts had gone off the air, Richard Nixon appeared in a live telecast to address the nation about the previous week's Supreme Court decision that ordered him to turn over the Oval Office tapes.

All that day, from the time the televised speech had been scheduled early in the morning, people in Washington and in the newsrooms had been buzzing about what Nixon might do or say. The Watergate scandal had gone from being a dry story about politics to a true human drama, even with touches of a soap opera. Millions of people were waiting to hear what the president was going to do next.

"I do not intend to resign," he said early in the broadcast. "Your president is not a quitter—you have a right to know that. I intend to finish my term of office, which a majority of American voters have given to me.

"You know, I was on the track team back in Whittier, California, and I have to tell you that I wasn't very good. But no matter how far behind I fell in any of my races, I never stepped to the side. I never gave up. And I do not intend to give up now."

At the White House, a small group of speechwriters had gathered to watch the speech on a television in David

Gergen's office. Buchanan, Ben Stein, and Gergen were exasperated when Nixon began telling his old story about his days on the Whittier track team. He was to give a historic, carefully crafted speech on the power of the presidency, and a digression about his high school loser days in Yorba Yahoo, California, wasn't part of the script.

"And the primary reason that I will not resign is *not* because I think Richard Milhous Nixon is indispensable to this great nation of ours—heh, heh—but because I believe that one branch of government should not be beholden to another."

In the White House, the speechwriters were pleased that Nixon was back on track and had even managed not to stumble awkwardly over the deliberately placed [CHUCKLE].

"Although I have certainly disagreed strongly with many of the Supreme Court decisions over the past twenty years . . ." (This was a carefully scripted and not altogether subtle attempt to gain support in the Southern states.) ". . . I would never call or write any of the justices to try to influence how they might vote. Likewise, I do not believe that the court has the authority to instruct the president in how he should conduct the affairs of his office. I believe that such an action is an unconstitutional violation of the checks-and-balances system that our forefathers set out almost two hundred years ago. And so I shall not comply with the Court's recent decision regarding this office. Though I have lost my base of political support, I will fight for this principle. If need be, I will appear in the Senate in person and will carry this struggle to the final conclusion.

"There are legitimate national security discussions on the tapes in question that preclude their release to the public. I do want to see this investigation completed fully, however, and so I will release transcripts of the tapes as soon as my staff and the proper staff at the National Security Agency have had an opportunity to remove any material that might potentially damage national security."

Phase two of the speech (which had been given the title "Long Bomb" to please the First Football Fan) was to be of a more personal, connecting nature. "I have made errors," Nixon said, trying to stare holes into the lens of the television camera. "Too many of these errors were because I was too loyal to members of my staff, even if some of these staff members conducted themselves in a manner that some people might consider illegal. These people have all resigned or been fired from their White House duties, and although some of the lapses in conduct were serious, I believe that the ruination of their promising careers is punishment enough.

"And so, I will tell you tonight, that when this investigation is completed, or on my last day in office, whichever comes first, I will grant pardons to these men for their time of service to the president. I have decided on the delay because to do so sooner would be to interfere with the workings of the congressional investigative process."

Back in 1952, when Nixon had given his Checkers speech as a candidate for vice president, he had outflanked the World War II general Dwight Eisenhower, who wanted to dump him from the ticket, by urging his supporters to write to the Republican National Committee to insist that he be kept. On this night, Nixon supporters were already distributing blank postcards preprinted with the addresses of their local representatives and senators. At the end of the speech, Nixon asked voters to write to their congressmen and tell them how they felt about the damage that partisan attacks were doing to the presidency.

The speech lacked the melodrama of a cocker spaniel puppy that was about to be made homeless, which the Checkers speech offered, but the next day several wags nicknamed Nixon's effort the "Checkmate" speech for the crafty political moves Nixon had performed. For one thing, if there was information on the tapes showing that Nixon had been involved in the cover-up from the beginning—and, of course, there was—Nixon had nothing to lose by refusing to turn over the tapes to the investigators. He might be impeached for defying the Supreme Court, but once people heard what was on the tapes and discovered that he had been lying for the past year, he was certain to be impeached anyway. At least this way the Senate would lack conclusive evidence, and there was a chance that it would vote to acquit him.

Also, by offering pardons to the men involved in Watergate, he was buying their silence. This was perhaps too obvious in his speech, but it was necessary. With the prospect of prison facing them, there was no doubt that one or two or all of them would have torn Nixon to shreds with their testimony. By the time they were finished talking to the senators, Nixon knew, they not only would have placed him at the Watergate Hotel on the night of the break-in, driving the get-away car, they would have had him on the grassy knoll in Dallas in 1963 and in Alger Hiss's garden placing microfilm in a pumpkin in 1950. Now their best chance to avoid wearing prison blues was to help Dick Nixon survive.

But Nixon's smartest move was defying the Supreme Court, which had little actual power to force him to comply with its ruling. Alexander Hamilton had written in

Federalist Paper No. 78 that the court was the "least dangerous branch," which had "neither force nor will" to enforce its decisions. And Nixon would soon expose that flaw. As the weeks went by and the promised transcripts of the tapes were nowhere to be seen, the Supreme Court could do little to force Nixon to hand over the tapes. What could they do, send in the U.S. Marshals to find the tapes? Under whose authority? The U.S. Marshals, the FBI, the Secret Service, all were under the executive branch. As soon as Nixon ordered them off the White House grounds, they would have to comply.

The next week the House began debating the three articles of impeachment. The representatives heard dozens of witnesses, and there was strong circumstantial evidence that the president himself had participated in the cover-up of the Watergate crimes. There was no "smoking pistol," as it was so often termed at the time—no direct evidence linking Nixon with the illegal activities. And without the White House tapes, there never would be.

Two weeks later, on Friday, August 23, the House voted on the three articles. Article One, the obstruction of justice charge, passed the House by a vote of 300 for impeachment and 135 opposed. Article Two, abuse of presidential power, passed 304 to 131. Article Three, defying subpoenas, passed 248 to 187. President Nixon became the second U.S. president to be impeached, joining the hapless Andrew Johnson in that small club.

At the White House, it was a somber day as the staff watched the House vote live on television. The Nixon family had gathered in their private quarters, awaiting word of the inevitable outcome. Nixon had been fatalistic about the House vote. He knew that he didn't have enough time to counter the wave of anger that had washed through the chamber after his Checkmate speech. But now there was a new battle, the trial in the Senate, and this was a battle that Nixon thought that he had a chance to win. "Now it's a new ballgame," Nixon told his three closest advisors, Haig, Kissinger, and Buchanan, that evening. "Those liberal fascists think they can run us out of town. Well, those bastards don't understand presidential power. They don't know what a person in this position can do if it is necessary. Look at Lincoln and what he did during the Civil War. Look at FDR during the Depression. The president can use a different set of rules if there is a crisis, and gentlemen, I think anyone would consider this to be a crisis."

In volume four of his autobiography, *Years of Remembrance: Other Times I Saved the World, But Forgot to Mention*, Henry Kissinger recalled that he had responded, "We must be very careful that we don't do anything that might alarm other nations or endanger the reputation of the presidency." But during a later investigation of the events of September, which came during the Carter administration, another taping system was discovered that revealed that Kissinger's actual response had been, "We must be very careful that we don't do anything to alarm the Soviets, but this is a situation that they will understand, and they will admire your courage in standing up to your opponents, Mr. President, as we all do."

Buchanan was ready to run to the ramparts to defend Nixon, armed with FBI files on several of the senators. "We've got the goods on those mother*******," Buchanan said. "Now it's time to hold their feet to the fire. And it's going to be a furnace."

But it was Haig, known for his impulsive temper, who became nervous at the direction Nixon was headed.

After an unusually short Labor Day break (on this occasion Congress gave itself only two extra days off, in addition to the usual holiday, instead of the customary four weeks), the Senate began considering its first impeachment trial in 106 years. Senate majority leader Mike Mansfield had begun preparations for a Senate trial early in the summer when it appeared that it might become necessary, and his aides had even worked out schedules with the television networks by the time the Supreme Court handed down its unanimous decision in late July. Senator Robert Byrd was not in doubt about the outcome: "We've got seventy votes for conviction; we need to get this unpleasantness over with quickly." But many other senators didn't agree: there was a strong sentiment that if the Senate were going to remove a president from office—a president who just two years before had enjoyed the largest landslide victory of any U.S. presidential candidate up to that time—the case against him needed to be clear and convincing. Throughout Washington, that phrase was heard over and over by senators from both parties. The evidence needed to be "clear and convincing," the testimony needed to be "clear and convincing," the entire Senate trial needed to be "clear and convincing."

After two weeks of such discussions and votes on the rules of the impeachment trial, Chief Justice Warren Burger swore in the senators as jurors on September, 17, 1974. The senators had planned to call more than fifty witnesses in an attempt to make the expected guilty verdict appear "clear and convincing." But by the time the day's events were over, it was President Nixon, and not the U.S. Senate, who gave the conflict its clarity.

The day before, as Nixon was crossing an alley between the White House and the Executive Office Building, a crowd of protestors had begun screaming at him and shaking their fists. The crowd, which numbered more than a hundred, surged toward the president, and Nixon realized that the half-dozen Secret Service agents would be powerless to stop them if the crowd intended harm. The episode reminded Nixon of his visit to Caracas, Venezuela, when he was vice president. In a riot that occurred during his trip, the Venezuelan government had had to clear a corridor to the airport for the vice president, using tanks and tear gas. This time there were no tanks or tear gas to be had, Nixon knew.

That afternoon Nixon told Haig about what had grown in his mind into a riotous mob that had tried to attack him. "No president since Lincoln has faced this kind of hatred," Nixon said.

Haig agreed. "This wouldn't be a tough place to take by force," Haig said, looking around the Oval Office. "This place isn't a fort, after all. F****** cars drive right by on Pennsylvania Avenue. You could roll a truck full of . . ."

"What kind of help might be available, in a situation?" Nixon interrupted. Haig knew that Nixon already knew what troops were available; this was just Nixon's way of gauging Haig's reaction to calling troops to the White House.

"The quickest would be the marines at the honor guard barracks here in Washington. Then there's the officer training facility at Quantico, Virginia—that's about thirty miles away. After that, you would be talking about the Eighty-second Airborne in Fort Bragg, North Carolina."

"We wouldn't need to get James Schlesinger over here for that, right?" Nixon asked about the secretary of defense. It was a shame to see the man act like this, Haig thought. The old Nixon didn't need constant reassurance from his aides about his decisions. During the Vietnam War, Nixon hadn't bothered to consult Melvin Laird, the secretary of defense—he had simply given orders to the generals himself. Now he's second-guessing his every step, Haig thought.

"There's nothing that requires that. You could call the marine commandant, Cushman, yourself—after all, you're the commander in chief!" Haig knew that this was all familiar to Nixon already. Nixon and Robert Cushman had a long history together. Cushman had been Eisenhower's military aide while Nixon was vice president. When the Nixon White House had wanted the CIA to handle a few illegal investigations in 1971, it was Cushman, who was the deputy director of the agency, who had authorized Gordon Liddy and Howard Hunt to perform the chores. Nixon later appointed Cushman head of the Marine Corps. If Richard Nixon needed help from the military, Haig thought, he couldn't ask for a better friend than Robert Cushman. If Nixon is asking me about this now, Haig thought, he's already run it through his head a hundred times.

At the same time that the House managers (who serve as prosecutors in the Senate) were being introduced and making their opening statements, a mile down Pennsylvania Avenue Nixon had given an order that—for his own safety and security, he insisted—he was calling up marines from their barracks in Washington and from

Quantico to protect the White House grounds. As word of his act reached Congress, Chief Justice Burger was forced to take a recess when he was unable to bring the senators to order.

After an evening dinner break, senators were rushed from cloakroom to cloakroom without regard to party to discuss the situation. Already the television networks had shown camouflaged marines standing behind the iron White House fence with M-16s at the ready. The Senate was about to begin an evening session that had been scheduled in prime time to accommodate the television networks, when Mansfield was given an urgent message from Al Haig. Mansfield asked Burger for a recess and left the Senate floor as the senators again tried to make sense of the day's events. Fifteen minutes later, Mansfield walked back into the Senate chamber, accompanied, to the great surprise of all the senators, by Vice President Gerald Ford. Mansfield was recognized by Burger, and then he turned and addressed the senators from his seat. "I have just received information that the president has mobilized the Eighty-second Airborne Division from Fort Bragg, North Carolina, with the apparent mission of . . ." Mansfield looked down at a piece of scrap paper in his palm, "of 'providing security to Washington and the federal government.' I have spoken with the president's chief of staff, Mr. Haig, and he has informed me that the president has instructed that no further information be given to anyone, including the Joint Chiefs of Staff, about the soldiers' mission.

"Mr. Chief Justice, in light of these events, I move that after a ten-minute recess, we conduct a roll call vote on the guilt or innocence of Richard Nixon."

During the recess, Mansfield and several other senators surrounded Ford to discuss what to do after he was sworn in as president. Ford insisted that the Joint Chiefs of Staff be standing by via telephone from the Pentagon. "I want that phone right there," Ford said, pointing to Mansfield's table at the front of the chamber. "As soon as I drop my arm I want to be on the phone with the Chiefs."

Several of the senators expressed concern about the nuclear "football," the briefcase containing the codes that would allow the president to launch a nuclear attack. Others wondered aloud if Nixon was acting spontaneously, or if this was just the beginning of some plot to seize control of the government that had been prepared months earlier. If that were the case, then he must have arranged for at least part of the military to support him even if Congress voted for his conviction and removal. Adding to the worry was the reality that there were no procedures put down on how exactly a convicted president would be removed if he did not willingly leave the White House. "It may be necessary for you to go down there, to the White House," Mansfield told Ford. "We need to see what other troops we have available in case he's able to surround the White House. We might need to go in there and remove them."

"God damn it, I'm not going to go down to Lafayette Park and climb on top of a tank and order Nixon to leave the White House," Ford said. "We have to get control of this some other way. The United States of America isn't going to be governed at the end of a gun!"

Out of nervousness, Chief Justice Burger began pounding his gavel to call the Senate jury to order a minute early, but the room quickly became silent. Burger ordered that the vote be taken, and as the clerk called out the names of the senators, all appeared to be voting for conviction. Earlier in the day Nixon could have counted on as many as two dozen of the one hundred senators to vote for acquittal, but now ninety-seven senators sharply called out guilty, and only three voted for acquittal. (The three senators who voted for acquittal all claimed later that they had done so because they could not vote to convict on rumor, but nonetheless, all three lost their seats when they next faced reelection.)

As soon as the vote was taken and Burger announced the verdict, Ford rose from a chair that had been placed at the side of the chamber, walked over to the chief justice, and was sworn in as president at 11:50 P.M. At that moment, the members of the Joint Chiefs of Staff, working under orders prearranged just moments before, issued new orders. The marines on the White House lawn were ordered to muster in Lafayette Park, across the street from the White House, and to await further orders. The soldiers from the Eighty-second Airborne were ordered to continue to Washington because of concerns that Nixon may have made some sort of arrangement with a rogue military commander. It was an unnecessary precaution. At half past midnight, the black presidential limousine passed through the White House gate headed for Andrews Air Force base. From there former president Nixon and his family were flown back to their home in San Clemente, California, in the middle of the night.

Ford did arrive at the White House at 2 A.M., but he did so only to reassure the staff there that the crisis was over. He declined to spend the night in the White House, but instead returned to his home. The next morning at 7:30 A.M., President Ford entered the Oval Office to begin his term as president.

Nixon released the White House tapes to the public on August 5, 1974, and resigned four days later. In that last week in the White House, Nixon had talked about several alternatives to resignation. The most frightening of these were discussions he held with a few people about mobilizing the Eighty-second Airborne. What Nixon had in mind will never be known, but several people were alarmed at the fact that he even mentioned such a possibility. When Haig told Kissinger that Nixon was considering mobilizing the Eighty-second Airborne to protect the White House, Kissinger said that this was "nonsense," telling Haig that a presidency could not be conducted within a circle of bayonets. "We were living in a surrealistic world," Kissinger said later. "Its victims had coexisted with a nightmare for so long that it had come to be the natural state of affairs. They had reduced their peril to a banality and therefore could not believe in its culmination."

Leon Jaworski, the special prosecutor, had suggested to the grand jury hearing the evidence against Nixon that it would be better to impeach the president than to indict him in a criminal court. " 'What happens if he surrounded the White House with his armed forces,' " a member of the jury recalled Jaworski saying. " 'Would the courts be able to act?' "

Of course, in the end, Nixon did not go on television to fight with the Supreme Court, he did not put the nation through the ordeal and expense of a Senate trial, and, although it seems fantastic to think about now, he did not circle the

White House with troops. He simply resigned, walking away from the office that had defined his life's quest for twenty-five years.

"You've got to say this for him—he had respect for the government, because he stepped out [of the presidency]," one of the original Watergate prosecutors told journalist Seymour Hersh. "If he were a Hitler or a Stalin, he'd have gone all the way, brought the house down. And that's what Jaworski was afraid of and that's what we were afraid of."

Chapter 26

IBM Steamrolls Microsoft

☙

In one of the worst-timed business decisions in U.S. history, computer giant IBM contracted in 1980 with a small company in the suburbs of Seattle to create its new personal computer's operating system. What if IBM had decided to write the software in-house instead?

THERE is no reason for any individual to have a computer in their home," proclaimed Ken Olsen, president of Digital Equipment Corporation, at the World Future Society convention in Boston in 1977.

One person who wasn't listening was Bill Gates. His vision was exactly opposite that of the president of Digital. Gates had made a prediction of his own: He believed that every home in the industrialized world would someday have a computer. Gates's prediction had a fanciful 1960s World's Fair absurdity, like the predictions that by "the year 2000" everyone would be flying to work in personal aircraft. But Gates was so certain of his prediction that he decided to leave Harvard at age nineteen to start his own company.

In 1977, the year of Olsen's prediction, the first microcomputers arrived on the market. Apple Computer, which had begun as two Berkeley students and assorted friends working in a garage, had released its first computer, the Apple II. Radio Shack was Apple's chief competitor, with its Tandy computers, but Atari and Commodore also released computers.

In 1980, microcomputers were still little more than toys for playing arcade games. Although these small computers were getting hardly any notice from the public at large or the business community, the world's largest computer company, IBM, decided that it wanted to build a small, personal-sized computer that could appeal both to small businesses and home users. A project team had been working on such a machine for two years, but the computer, called the Datamaster, wasn't ready for release to the market. In July 1980, Bill Lowe, the director of IBM's technical laboratory in Boca Raton, Florida, was in a meeting with the chairman of the board, Frank Cary, who complained that IBM wouldn't be able to develop a small computer without hundreds of people spending years of effort. By the time IBM came to the party, it would be all over except for the cleanup. "No sir, you're wrong," Lowe told Cary. "We can get a project out in a year."

What Lowe had in mind was to assemble parts and software from other suppliers and encase it all in a box with IBM's name on it. Almost immediately Lowe was put in charge of Project Chess, with the goal of building IBM's first microcomputer by the summer of 1981.

For most of the twentieth century, IBM was nearly synonymous with computers. IBM had begun as the Calculating-

Tabulating-Recording (CTR) corporation in 1910, a merger of three office machine companies. Shortly after the company was formed, it hired Thomas Watson as its president. Watson had previously worked for one of America's greatest business success stories, National Cash Register, or NCR. But while working for NCR, Watson had violated the Sherman Anti-Trust Act and was forced to leave the company. Watson soon turned CTR, which was renamed International Business Machines in 1924, into a company even bigger than NCR. Under Watson, IBM dropped many of its products to concentrate on mechanical punch card tabulators, which were in high demand. By the 1930s, IBM was one of the most profitable companies in the nation.

The company was so successful, in fact, that the U.S. government sued it in 1932, accusing it of violating antitrust laws. Even while it was under investigation, however, IBM continued to grow, fueled in a large part by the U.S. government, which had a need for tabulators in the newly created Social Security Administration.

During World War II, IBM helped to build one of the world's first computers, the Electronic Numerical Integrator and Calculator, or ENIAC. ENIAC was a room-sized behemoth that contained five hundred miles of wire and could only handle calculations that would not tax a modern calculator. Unable to imagine the technological breakthroughs to come, IBM president Watson announced in 1943, "I think there is a world market for about five computers."

But insurance companies, government agencies, and large corporations had a need for computing power, and after World War II the sales of IBM computers grew, especially after transistors replaced vacuum tubes in the late 1950s. A half-dozen companies sold the expensive, mammoth machines—companies such as General Electric, RCA, Honeywell, and NCR—but in 1968, IBM still held 65 percent of the nation's computer market.

Again, this success did not go unaccompanied by charges of illegal activity. Rival companies charged that IBM had its customers place phantom orders for computers that were out of stock to prevent them from buying from competitors. By the end of the decade, IBM was hit with nearly two dozen separate antitrust lawsuits.

Microsoft didn't have an American business pedigree like IBM. The company began in 1975, when Bill Gates entered the computer business in his Harvard dorm room. Gates and his friend Paul Allen had been computing prodigies, writing their first computer program while they were attending a private prep school. The program scheduled classes for the school, but it had a hidden bit of code that made sure that the boys were put in classes with the best-looking girls. When he was a senior in high school, Gates was already working full time for TRW as a computer programmer. Gates left for Harvard with the goal of creating "the IBM of software."

Gates and Allen had written a software language for BASIC, and while attending Harvard, the two approached the manufacturer of one of the first microcomputers, the MITS Altair, about including BASIC on the machine. Even though he was still a teenager, in this deal Gates, whose father was an attorney, produced the creative language that would change the business world and make Gates and several of his friends billionaires. It wasn't computer code but a legal contract.

According to the contract Gates had convinced MITS to sign, the computer manufacturer could install and sell BASIC, but Microsoft kept the full rights to the software and was free to sell it to other companies. It was as if a best-selling author was free to sell his manuscript to four or five publishers, all of which put books out on the market at the same time. With publishers competing and publicizing the books, trying to outdo one another, the author could sit back and enjoy the fruits of the competition. This is what Microsoft did as it wrote other computer languages, such as FORTRAN, and sold the rights to use the languages to such well-known companies as General Electric, NCR, Zenith, Sharp, Texas Instruments, Radio Shack, and Apple.

But what if the best-selling author doesn't have a manuscript to sell? He could always buy one from a less well-known writer and put his name on it. That's what Bill Gates and Paul Allen did when IBM, in the guise of opportunity, came knocking at Microsoft's door.

IBM was interested in buying an operating system for a microcomputer called CP/M, and the IBM managers were under the mistaken impression that Microsoft had written the software. Gates told IBM that it was actually produced by a small company in Pacific Grove, California, called Digital Research, and he arranged for a meeting with the owners of that company and IBM.

On the day that IBM representatives arrived at the company, which was located in an old Victorian house, they discovered that the founder, Gary Kindall, had left to go flying in his private airplane. The IBM managers tried to begin the meeting without him, but Kindall's wife refused to sign the ridiculously one-sided nondisclosure statement that had been shoved in front of her. IBM didn't discuss its business plans with potential business partners unless they signed the agreement, and so the IBM managers left without telling the Kindalls what they had come for.

IBM still needed an operating system for its new microcomputer, and representatives of Big Blue invited the Microsoft crew to their offices in Boca Raton to discuss buying one. One IBM engineer said that Gates arrived at the meeting looking like "a kid that had chased somebody around the block and stolen a suit off of him, and the suit was way too big." But Gates was impressive, not only for his brilliant knowledge of computers but also for his obvious business savvy. At the meeting, Gates promised to provide an operating system for the computer. Microsoft didn't have an operating system, but the Microsoft people thought they knew where they could get one (information that wasn't revealed to IBM, of course).

Microsoft co-founder Paul Allen knew of an engineer named Tim Patterson, who worked for a small business back in Washington called Seattle Computer Products. Aggravated by delays in receiving copies of CP/M, Patterson had simply written his own operating system software. The new software, called QDOS for "Quick and Dirty Operating System," was based on CP/M and involved what Patterson called "low-level borrowing." Gates and Allen thought this could work as the operating system that IBM was looking for.

Just two days after Ronald Reagan defeated Jimmy Carter in the 1980 election, Bill Gates signed his name to the contract with IBM. Perhaps awed by the size of IBM compared with his small company, Gates didn't use his title of president on the contract, instead signing "partner."

The relationship between IBM and Microsoft was a contentious one from the start. IBM was famous for the 1950s-style dress code that insisted that its employees wear suits, navy or gray; white shirts; and ties, a uniform that earned the company the nickname "Big Blue." Microsoft was the opposite, with no dress code at all. On the day that he had signed the contract with IBM, the unshaven Gates was wearing a ratty sweatshirt. In Seattle the Microsoft employees favored informal dress such as flannel shirts and jeans; it wasn't long before the culture clash between Big Blue and Big Plaid was on.

IBM had a Cold War militaristic attitude toward company secrets, and that attitude extended to any contract work that Microsoft was doing for the company. IBM insisted that Microsoft's programmers work in a windowless room that was to be locked at all times, whether someone was working inside or not. The only room that Microsoft had that fit the requirement was a storage closet, and so the programmers and computers were stuffed inside the cramped, hot, and soon-to-be-odorous room. But the Microsoft staffers were able to hide their slack security from the visitors from Big Blue, and work continued on the operating system and other software for IBM's new computer.

But Gates still didn't own the operating system software IBM was counting on. The day after Gates signed the agreement with IBM, a representative of Seattle Computer Products sent Microsoft a proposal for the sale of the software, which Gates had already promised to deliver to IBM. The two companies negotiated for months until on July 27, 1981—just three weeks before the IBM PC was introduced—Microsoft finally bought all rights to the 86-DOS software, which they renamed MS-DOS, for $75,000.

In August 1981, the IBM PC was unveiled. Bill Gates didn't attend the news conference—he had asked to be invited to the release announcement, but his request had been denied. The IBM PC became not just successful, but a paradigm-shaking icon. Soon other companies joined in. Zenith produced a computer with two processors so that it could run software designed for both 8-bit and 16-bit computers. Hewlett-Packard designed a computer with a touchscreen so that users could point to the file they wanted to work on. And Wang was producing a successful computer that was a dedicated word processor. But none of these computers were nearly as successful as the capable IBM PC, and by 1983, three million microcomputers were in use, most of them IBM PCs. And on every PC was a copy of MS-DOS.

IBM paid Microsoft a total of $700,000 to adapt several computer languages and to develop several new programs for the new IBM computer. But in addition to the upfront money, Microsoft was to get a royalty payment for each copy of each of the various pieces of software that was sold by IBM. IBM didn't buy MS-DOS, but merely licensed its use on its computers. Microsoft was free to sell the operating system to other computer manufacturers.

That wasn't a problem as far as IBM was concerned. To properly run MS-DOS, a computer would have to have a piece of hardware, a ROM-BIOS chip, which only IBM manufactured. Microsoft could sell all the copies of MS-DOS it wanted, but only an IBM PC could run it. But within a year, a new computer company, Compaq Computer, spent $1 million to reverse-engineer the ROM-BIOS chip, and within two years after the IBM PC debuted, the Compaq computer was on the market. It was an immediate success. Nearly fifty thousand were sold in its first year. Soon other companies began selling the ROM-BIOS black boxes to anyone who wanted to build an IBM clone, so that within a few years resourceful undergraduates began building IBM clones in their dorm rooms.

Apple Computer didn't sit on the sidelines and watch IBM steal the microcomputer crown. In 1984 Apple released the

Macintosh computer. Although most people focused on the computer's stylish design, the operating system was the real breakthrough. MS-DOS machines used numbers, words, and hundreds of obscure codes to work. The Macintosh instead used symbols designed to look like items found in an office, such as file folders and trashcans. But the Macintosh didn't sell as well as predicted, and within two years after its release, Steve Jobs, cofounder of Apple and the visionary behind the Macintosh, was forced out of the company.

Moore's Law, the computing cliché that says that processor speed doubles every eighteen months, soon became a problem for IBM. Big Blue spent at least three years to bring new products to the market, but with new processors coming out within months—such as with the introduction of the 386 processor in 1986—IBM found itself selling obsolete machines. By 1990, IBM revenues actually fell for the first time in fifty years, and IBM began laying off employees. That year, Bill Gates predicted that the company would fold by 1997.

While IBM's revenues were falling, Microsoft's were soaring. Microsoft offered public shares of stock for the first time in March 1986 for $25.75 a share. Within a year the stock had tripled, and Bill Gates, the largest stockholder, was worth more than a billion dollars. He was thirty-one years old at the time.

In 1984, *Fortune* magazine declared, "IBM sets standards in this industry, not Bill Gates." But IBM had allowed Microsoft to decide what its computer's operating systems would look like, what functions they could perform, and what software they could run. Even in 1984, IBM may have been setting the standards, but they were licensing them from Bill Gates. *Fortune* declared that IBM would soon write its own operating system for the PC and that IBM "would probably wean customers away from MS-DOS in a gradual series of steps." But unfortunately for IBM, that thoroughbred had left the stable with Bill Gates riding. It was too late to close the barn door.

WHAT IF?

What if IBM had decided to write the operating system for its personal computer in-house instead of licensing it from Microsoft?

At the Microsoft offices in Bellevue, Washington, Bill Gates sat on the floor of his office, rocking back and forth, his head down and his arms folded across his chest. "What do we do? What do we do now?" he moaned as other Microsoft employees stood by helplessly. Suddenly Gates's demeanor changed, and he stood with his face flush red. "They can't do this! They just can't leave us like this!" He picked up a lamp from his desk and threw it to the floor, where it bounced instead of breaking with a gratifying crash. He then swept his arm across the desk, sending the papers and desk supplies flying across the room.

Gates held a letter in his hand from IBM informing him that the relationship was over. "That's a lie!" he shouted, pointing at the letter. "I challenge their facts! I've never heard of such a thing. It's all nonsense!"

Although Gates's tantrum continued for the next two days, there was nothing he or Microsoft's attorneys could do about IBM's decision. The contract had specified predetermined points at which IBM could withdraw from the agreement, and they had decided in March 1981 to exercise this clause in the contract.

During their visits to Microsoft, the IBM managers

had been appalled at how little regard Microsoft had shown for the security of their top-secret project. Parts of the prototype computer had been seen in several parts of the office in full view, and papers from IBM had been spotted on a receptionist's desk. In more than one visit to Microsoft, IBM managers had seen that the door to the room where the PC's operating system was being developed was wide open, and once a Microsoft programmer had been spotted working on the prototype as a young woman, who appeared to be his girlfriend, sat nearby. "It's like we've turned our most important project over to a bunch of guys at a fraternity house," one IBM manager grumbled. "Who are his other customers? We don't know. Anybody could be walking through the building looking at what we're up to."

But the loose security wasn't the primary reason why IBM had suddenly withdrawn from the project. IBM software engineers had been comparing the QDOS operating system software that Microsoft was working on with the CP/M and found some alarming similarities. For example, hitting the function code 6 key resulted in a dollar sign in both systems. And to move a file in CP/M, a user typed in "PIP B: A: FILE," where in QDOS the command to move a file was "COPY A: FILE B:"—too close for comfort. IBM was in the midst of a Justice Department antitrust investigation, and IBM itself was suing the Japanese computer company Fujitsu, alleging that Fujitsu had stolen parts of a code from an IBM operating system. When the IBM managers showed the similarities to the lawyers at IBM, the legal eagles began screaming to anyone who would listen that IBM had to end the deal with Microsoft.

The divorce of IBM and Microsoft had been orchestrated by IBM programmers who had been working on their own code for the operating system for two years before Microsoft arrived in Boca Raton. These blue-suited lifers had been horrified at the thought of what this success by Microsoft might do to their careers at IBM. If a group of college kids in Seattle could do in a few months what a room full of IBM professionals hadn't been able to accomplish in two years, they would all soon be looking for jobs down at the local Radio Shack. Inspired by desperation, the IBM programmers had been able to make great strides in fixing the problems with their own operating system code, and they had discovered the deadly similarities between the Microsoft code and that of Digital Research.

The IBM PC encountered several delays, but the computer was finally unveiled in November 1981. It was an immediate success; IBM sold ten times as many computers in the first three years as it had expected. By the time Apple Computer was able to counter with the Macintosh computer in 1984, it was too late. Most of the software on the market was written to run on the IBM machine, and the beige IBM box saw only increasing sales.

Microsoft quickly adapted to the new landscape and wrote several successful pieces of software for the IBM machine. But it was the Microsoft collection of business software for the Macintosh, released in 1985, that finally caused Apple sales to rebound and IBM sales to level off. Microsoft soon found itself a dominating force in computer software. The microcomputer market, which had been teeming with dozens of small computers each with its own strengths and weaknesses, had narrowed to just two main competitors, IBM and Apple.

Although Apple Computer had a successful product in the Macintosh, the computer wasn't welcomed by the business community as the revolutionary product that had been anticipated. In 1985, John Sculley, the president of Apple who had been recruited from Pepsi Cola by Apple CEO and spiritual leader Steve Jobs, decided that it was the mercurial Jobs who was preventing the company from stopping the IBM juggernaut. In a move that created headlines around the nation, Apple fired its CEO and cofounder.

Apple's board of directors knew that the company needed someone who was brilliant both at business and with computers. After six months of rumors about who the new CEO of Apple would be, in February 1986 Apple presented its new CEO, William Gates III.

Apple had bought out Microsoft (although the stubborn Gates had insisted that the deal be announced as a "merger") and installed the CEO of Microsoft as the leader of the nation's second-largest computer hardware and software company. Gates appeared at a press conference in his usual tan slacks, sweater, and open-collared shirt and began answering questions from the media with a goofy grin on his face. "Computers will soon be part of a multimedia continuum," Gates said, his voice characteristically squeaking, a trait that seemed to place him in permanent adolescence. "As these products become available, and as Apple makes these things easier to use, more and more people will be using computers, and Apple will be leading the way."

Gates (along with former Microsoft president Steve Ballmer, whom Gates had brought with him from Microsoft) immediately began making the kinds of improvements in Apple's corporate structure and direction that had made Microsoft the darling of the business elite. Apple's stock began to soar as the investors on Wall Street bet money on Gates's leadership of Apple.

Then, in September 1986, Apple Computer dropped a bomb on the computer industry and it wasn't the second-generation Macintosh. Gates directed his legal staff to deliver a letter to the U.S. Justice Department's Antitrust Division accusing IBM of anticompetitive business practices. The letter accompanied a report prepared by Apple's legal staff that accused IBM of every illegal activity shy of tying up programmers and spiriting them away in the night.

"It's well established that competition is required for the efficiency of marketplaces," Gates said in an interview on CNN. "You have to have procompetitive stuff. This all assumes that you like capitalism." Gates's invitation to the federal government to look into the computer industry wasn't welcome by many in that business who preferred that bureaucrats stay in Washington, D.C., and computer people live on the West Coast and never the twain shall meet. But Apple soon found approval from the general public, who viewed Apple as the small, scrappy company fighting a large, unethical Goliath. The antitrust complaint against IBM turned into an unexpected public relations bonanza.

The federal government responded to Gates's invitation by launching a new antitrust investigation of IBM in January 1987, exactly five years after it had dropped its thirteen-year-long antitrust investigation in 1982. Soon other companies began filing complaints about IBM's business practices with the Justice Department (DOJ) with such speed that the DOJ investigators weren't able to pursue every lead. After eighteen months of hearings, many of which featured the government's star accuser and witness, Bill Gates, a U.S. federal judge in August 1988 ordered IBM to separate its corporate divisions into two different companies, hardware and software. The day the verdict was announced, Bill Gates refused to comment, but one of Gates's admiring critics (a contradictory stance many business people took toward the Apple CEO) summed up the situation well. "Gates is Darwinian," he said. "If you strike him, he will find a way to eat you."

It wasn't the Sherman Anti-Trust Act but Moore's Law that broke up IBM. IBM typically took three years to bring a new product to market, but in a marketplace following Moore's Law, IBM computers were outdated soon after they were released. In late 1991, IBM president John Akers announced that each division of IBM would become its own business, able to make its own decisions, even if these decisions hurt the sales of another division. IBM would soon become a major producer of personal computers again, but the era of Big Blue dominance had come to an end.

Although Microsoft was a huge success, it was still smaller than computer hardware manufacturers. In 1995, Microsoft revenues were half those of Apple and one-fourteenth the size of IBM's. But producing software is much more profitable than producing hardware, and by 1999, ten thousand Microsoft employees had become millionaires thanks to stock options. Bill Gates was the world's richest man, with an estimated personal worth approaching $100 billion.

Personal computers were the peasant revolutionaries of the computing world. They assaulted the mainframe computers that dominated corporate America and replaced the centralized power with individual authority. In 1980, almost all the computing power in corporations existed in air-conditioned rooms housing giant mainframe computers. In 1987, 95 percent of the corporate computing power was housed in desktop PCs. Within five years, most of these personal computers would be linked via the Internet, and the power of this ability to move information around changed the way the nation conducted business.

Chapter 27

Dumping Dan

❧

Many voices clamored for George Bush to find a new running mate for the 1992 presidential election, but Bush remained loyal to Dan Quayle. What if Bush had selected a new vice presidential candidate?

MOST people in the nation were introduced to J. Danforth Quayle when the forty-one-year-old senator came bounding onto an outdoor stage in New Orleans, grabbed Republican presidential candidate George Bush's shoulders, and shouted, "Believe me, we will win because America cannot afford to lose! Let's go get 'em ! All right? You got it?"

Across the nation the electorate seemed to respond in unison: "Huh?"

From that moment on August 16, 1988, Dan Quayle became one of the most talked about political figures in America, largely for all the wrong reasons.

A few months before, in the midst of the news doldrums between the spring presidential primaries and the summer political conventions, the national press began speculating about what kind of president Republican candidate George Bush would make. Bush pointed out that the first major decision he would make for his administration would be the selection of his running mate: "Watch my vice presidential decision. That will tell all."

When Bush announced Dan Quayle as his choice for his running mate, the junior senator from Indiana was unfamiliar to most Americans, but to the horror of the Republicans, news about Quayle would dominate the headlines for the next two weeks, and nearly all of it was bad.

Even before the nominees left the Republican convention in New Orleans, Quayle was a liability for Bush. There were questions about how Quayle had avoided the Vietnam War draft by joining the National Guard; he had been able to get into the Guard because of his family's influence. There were reports of a sex-for-votes scandal on a 1981 golf trip that he had joined. Quayle's wife, Marilyn, dismissed the idea that Quayle had participated in a sexual quid pro quo by insisting, "Anyone who knows Dan Quayle knows that he would rather play golf than have sex any day." Quayle didn't help when Dan Rather, the anchor of the *CBS Evening News*, asked the candidate what his worst fear was, and Quayle answered "Paula Parkinson," who was the female lobbyist on the 1981 golf junket.

Quayle soon revealed a special talent for misstatements. Campaigning that fall, he described the Holocaust as "an obscene period in our nation's history." When others suggested that it was actually part of Germany's history, and not that of the United States, Quayle responded, "Our nation was on the side of justice. . . . I mean, we, we all lived in this century. I didn't live in this century, but in this century's history." Not surprisingly, the Quayle gaffe-watch became a national pastime, and everyone in the nation (outside of the Quayle family) was in on the joke.

The jokes weren't mortal wounds to the Bush campaign,

however, and the Bush-Quayle team easily defeated their Democratic challengers Michael Dukakis and Lloyd Bentsen by winning 54 percent of the popular vote. Despite the fact that the competitive atmosphere of the election had ended, the Quayle jokes continued, and many were mean-spirited. "The Secret Service is under orders that if Bush is shot, to shoot Quayle," Massachusetts senator John Kerry quipped.

During the Quayle's tenure, a vigorous cottage industry sprang up solely to make fun of the vice president. A newsletter, *The Quayle Quarterly*, tracked all of the veep's missteps, and a tongue-in-cheek organization called the "President's Prayer Club" began selling T-shirts with the motto, "Keep George Healthy." There was even a 1-900 telephone line where people could call for a daily joke about the vice president. By the end of Quayle's second year as vice president, the Center for Media and Public Affairs reported that more jokes had been made about Quayle on late-night talk shows in 1990 than any other person, beating out Iraqi dictator Saddam Hussein and Washington D.C. mayor Marion Barry.

It was probably the largest assault of ridicule that any American politician has been forced to endure.

The Quayle jokes subsided on August 2, 1990, when Iraq invaded the neighboring nation of Kuwait and began threatening an invasion of Saudi Arabia, endangering much of the world's oil supply. For the next several months, the nation watched as war with the world's fourth-largest army appeared more and more likely.

Immediately after the invasion of Kuwait, President Bush began putting together a coalition of nations to stop Saddam Hussein, and on August 8 he deployed the Eighty-second Airborne division to Saudi Arabia. As the buildup of American forces continued past the first of the year, critics of the Bush administration predicted that the U.S. faced a military disaster. "It'll be brutal and costly," warned Massachusetts senator Edward Kennedy. "The forty-five thousand body bags the Pentagon has sent to the region are all the evidence we need of the high price in lives and blood that we will have to pay."

The United States began an air bombardment of Iraq in January 1991, and when Hussein refused to withdraw from Kuwait, the U.S. and Allied forces invaded Iraq on February 24. The war was a complete rout of the Iraqi army, and after four days of fighting, President Bush announced a cease-fire. The United States and the Allied forces had stopped Hussein's aggression in the region after just one hundred hours of fighting, with a loss of 131 soldiers.

The Gulf War had been such an incredible success that Bush briefly enjoyed a 90 percent approval rating in public opinion polls—the highest numbers any president had seen since polls began gauging presidential popularity. As pundits looked toward the 1992 presidential election, they saw almost no chance that Bush could be defeated in his bid for reelection. One Democratic pollster moaned, "I don't know whether you'd be better off serving in the Confederate Army or being a Democratic presidential hopeful right now."

But while support for Bush was at historically high levels, Vice President Quayle continued to struggle for public approval. In the spring of 1991, after going jogging, President Bush experienced a heart problem. The damage to the president's heart was minor, but the nation's political psyche had a major infarction—the idea of a Quayle presidency caused renewed attacks on the beleaguered vice president. "A heart flutters, a nation shudders," wrote the *Chicago Tribune*. Even in Quayle's home region of Fort Wayne, Indiana, the *News Sen-*

tinel said, "Well, la-de-da-de-da, here we are just ambling along, average goofy Americans without a thought in our heads and . . . 'WHAT? PRESIDENT Quayle? OH GOD, PLEASE, NO, NO, NO!' "

After Bush's heart trouble, polls showed that only 19 percent of the nation considered Quayle fit for the presidency. "If 81 percent go for Bush and 19 percent go for me, then we've got just about everybody," Quayle responded. But calls for Bush to drop Quayle from the ticket were coming from around the nation. "Mr. Quayle is a sunny and unvengeful man. He is also weightier than given credit for, and might prove a surprisingly serviceable successor to Mr. Bush," said the *New York Times*. "But why not the best? Or at least, why not better?" More than fifty newspapers around the nation agreed with the *Times* and wrote editorials suggesting that Bush find a new presidential partner. *Time* magazine ran a cover with the line FIVE WHO COULD BE VICE PRESIDENT and included photos of Chairman of the Joint Chiefs of Staff General Colin Powell, Secretary of Defense Richard Cheney, Senator Nancy Kassebaum, and governors Pete Wilson of California and Carroll Campbell of South Carolina. *Newsweek* tagged along. On its cover was a photo of Quayle swinging a golf club, along with the caption THE QUALYE HANDICAP: IS HE A LIGHTWEIGHT — OR SMARTER THAN YOU THINK? *Newsweek* concluded, "Only Saddam is lower as a '92 VP choice."

But Bush was determined to stand by his Dan. "I've expressed my support for Dan Quayle," he said in response to a question about finding a new running mate for the 1992 election. "I think he's getting a bum rap in the press—pounding on him when he's doing a first-class job. And I don't know how many times I have to say it, but I'm not about to change my mind when I see his performance and know what he does."

But domestic problems soon changed the complexion of the 1992 presidential election. A recession that had begun in 1991 was deepening, causing many people to forget the support they had given the Bush-Quayle administration a year earlier. As people in the Bush administration tried their best to downplay the downturn, saying that an economic recovery was beginning (it was, as it turned out), Quayle stumbled along, *twice* predicting in speeches in California, "This president is going to lead us out of the recovery. It will happen." The next day, Quayle's motorcade passed a Burger King with a Help Wanted sign posted, and Quayle stopped the motorcade and pointed the sign out to the accompanying journalists, saying that the minimum-wage fast-food job was an "optimistic sign that things are turning around in California."

Although polls at the time showed that a majority of Americans wished that Bush would change his mind, and one out of four voters said that they wouldn't vote for Bush because of Quayle, there was still little reason for Bush to worry about his reelection. Bush continued his high position in the polls well into the election year of 1992. Even then it seemed as if he could coast to victory in the fall. His primary rival for the presidency, Arkansas governor Bill Clinton, had been shown to have lied to evade the draft during the Vietnam War, and despite his denials, evidence emerged that showed that he had cheated on his wife with an Arkansas newscaster-slash-nightclub singer named Gennifer Flowers. After Clinton secured the Democratic presidential nomination in the spring primaries, few people were giving him any chance of defeating Bush and Quayle. "He is a dead stone loser," one Democratic pollster said.

But by the summer of 1992, the wheels fell off the Bush bandwagon. Suddenly Bush found himself even with or behind his Democratic challenger in the polls. As the Republican campaign struggled along, Quayle continued to be ridiculed by the national media. In June, at a media opportunity at a spelling bee in New Jersey, a twelve-year-old boy correctly spelled "potato" by writing it on the chalkboard. As he began to return to his seat, Quayle insisted that he try again. "Add one little bit on the end," the vice president encouraged him. "Think of 'potato,' how's it spelled? You're right phonetically, but what else?" As the puzzled child added an "e" onto the end of the word, the vice president cheered. "There ya go! All right!"

The national media made the vice president's poor spelling skills a huge story, and the downtrodden veep had to endure weeks of more jokes on late-night television, this time focused on his inability to spell "potato."

For his part, Bush at times seemed oddly disengaged from the campaign. Immediately after the Gulf War, First Lady Barbara Bush had speculated to the media that her husband might retire and not seek a second term, and during the campaign Bush at times appeared to be a man who still hadn't made up his mind about whether he wanted to be reelected or not. He was spotted glancing at his watch in one of the presidential debates, as if he had another appointment to rush off to (he claimed later that he had been just checking to see how long his opponent had been answering the question), and in a press conference with a group of agricultural journalists, when a reporter from a farm publication asked Bush what he had done for rural America, Bush responded by talking about nuclear disarmament, citing it as an accomplishment that had benefited all Americans.

After a lackluster campaign, Bush lost his bid for reelection, with many Republicans jumping ship to vote for third-party candidate Ross Perot instead. In the final results, Bush pulled in 38 percent of the popular vote, Perot drew 19 percent, and Bill Clinton gained 43 percent and the presidency.

WHAT IF?

What if George Bush had chosen a different running mate for the 1992 election?

In June 1992, Vice President Dan Quayle surprised the nation by announcing, "I will not be a candidate for the vice presidential nomination." Over the electronic chirping of reporters punching in their office numbers on cell phones, Quayle continued: "This was my own decision, which I made after talking at length with my family. I have decided instead to enter the gubernatorial race for governor of my home state of Indiana."

As the reporters began shouting questions, Quayle— who had once described himself as "Dr. Spin"—attempted to explain the sudden change in plans. "I am from there, Marilyn is from there, and I would like my kids to experience growing up there, too. For too long my family and I have lived in the glare of the media spotlight here in Washington, and, quite frankly, we've suffered for it. I saw an opportunity to go back home and once again serve the people of my home state, and that's what I'm going to do.

"It was my own decision," Quayle repeated. "Just mine and Marilyn's, and after I talked on the phone with President Bush yesterday and told him of my plans, he offered nothing but his highest support and best wishes."

Despite Quayle's carefully laid-out reasoning for leaving the Bush administration, the national media saw through the smoke. DAN DUMPED! screamed the headline of the *New York Post* the following morning.

In fact, despite the concrete-solid denials of the Bush presidential campaign, Quayle *had* been dumped from the ticket. His days as vice president had become short from the time of Bush's less than strong performance in the spring Republican primaries, but it had taken vigorous machinations behind the scenes by Bush's top advisor, Secretary of State James Baker, to remove Quayle from the post he had clung to so tenaciously.

Baker had never held much respect for Quayle's professional abilities, nor had the two men ever warmed to one another personally. Baker had first tried to convince Bush to drop Quayle from the ticket after Bush's poor showing in the New Hampshire primary early in 1992, telling him that with Quayle on board he could lose the fall election. "If you lose the election, your success in Iraq and in Panama won't matter," Baker had predicted. "A one-term presidency is always considered a failure, no matter what transpired during that time."

Baker had been able to cite several precedents of vice presidents being asked to step aside. Bush had been a part of Gerald Ford's administration when Ford had dropped Vice President Nelson Rockefeller in place of Kansas senator Bob Dole; FDR had substituted Henry Wallace for John Nance Garner in '40, and even Abraham Lincoln had replaced his first vice president, Hannibal Hamlin, with Andrew Johnson in 1864. But despite the history lesson, Bush remained stubbornly loyal to Quayle. "Daaan Quaaaayle is up to the job, has a real handle on it. And he's learning more every day," Bush said. "I don't care if keeping him on puts us in deep doo-doo; I'm not gonna do it. Not gonna drop him like that."

But Baker and several other Bush cabinet officers, including Richard Cheney, the secretary of defense, and Lamar Alexander, secretary of education, were convinced that there was a possibility the Republicans could lose the fall election if something dramatic wasn't done to break the administration out of the doldrums that it had fallen into, and the most obvious solution to most people seemed to be to find a way to replace Quayle on the ticket. They had to find an alternative, one that was so attractive that it would overcome even Bush's stand. Baker, working with several members of the campaign committee, decided to approach the most popular person in the administration to serve with Bush. But Colin Powell wasn't interested.

Powell, chairman of the Joint Chiefs of Staff, had masterminded the military strategy that had overwhelmed the largest army in the Middle East, and following the Gulf War, he had scored approval ratings in the polls that were even higher than those of Bush. But while Bush had seen his poll numbers steadily decline to alarmingly low levels, Powell's had continued to float in the stratosphere. Powell had little interest in politics, and on two separate occasions he rejected any suggestion that he replace Quayle on the Republican ticket. After Bill Clinton won the Democratic presidential nomination in the primaries, however, Bush's friend Cheney approached Powell a third time. "I don't blame you for not wanting to jump into politics," Cheney said. "It would be hard on your family, and hard on you. You'd spend all you days being told where to go and what to say instead of giving orders. It's an unpleasant duty." Cheney paused, letting the word "duty" hang in the air. "But look at it from another point of view. If Bush should happen to lose the election—I don't think he will, but it is a possibility—then that draft dodger Bill Clinton will be in the White House as commander in chief."

Cheney paused again. Powell was a decorated veteran of the Vietnam War, and there was no need to ask him how he felt about someone who had schemed so strenuously to avoid serving his nation. Before Powell could respond, Cheney continued, "You created the Powell Doctrine: Go in with the force to do the job and do it right. No more fighting with a self-imposed disadvantage. The Powell Doctrine will change how America fights its battles for the next fifty years. Well, we need the Powell Doctrine here. We might win with the team we've got. But we might not. Why shouldn't we go into the fight with the best we have to offer?"

As Cheney sat across from Powell, there was no indication in Powell's face that anything he had said had moved the general's position even an inch. Powell stared at Cheney with a furrowed brow, looking quite displeased that this arrangement had been presented to him a third time. "Okay, I'll do it," he said, his terse response making it clear that he wasn't going to be an enthusiastic candidate.

With Powell agreeing to run with Bush on the fall ticket, the problem once again was Quayle. He could jump off the ticket or he could be pushed, but Baker knew that it would be easier for Quayle to make the jump if he had a soft place to land. After a late-night meeting between Clayton Yeuter, the chairman of the Republican National Committee, and Rex Early, the Indiana chairman of the Republican Party, a plan was organized.

Yeuter flew to Indiana to meet with that state's Republican candidate for governor, Linley Pearson. The Indiana gubernatorial candidate was curious about why the chairman of the RNC had such an urgent need to meet with him. He soon found out that it wasn't to congratulate him on his recent primary victory.

Yeuter told the Indiana candidate that he might lose the election to his Democratic opponent, Evan Bayh, the telegenic son of former Indiana senator Birch Bayh. It was going to be a difficult election. Bayh was extremely popular in the state, and any Republican challenger stood little chance of succeeding. Pearson wasn't convinced that defeat was certain, but Yeuter continued. If Pearson would step aside for the good of the party, in the next administration a person of his abilities would be a natural choice for a cabinet position, perhaps as commerce secretary. The Indiana candidate was smart enough to know that if he bucked the party on what was obviously its top priority, campaign donations from corporations and top Republican supporters would dry up like a creek in August, and he would lose anyway. The unspoken message was that he wasn't going to be the governor of Indiana, but he could choose whether to take the easy road or the rocky one. Two weeks after he won the Indiana primary, securing his nomination as the Republican candidate for governor, Pearson announced that he was withdrawing from the race for "health and personal reasons" and that the Indiana Republican committee would select a candidate.

The day before Pearson's announcement, Quayle was called to a meeting in the White House. Samuel Skinner, a golfing buddy of Quayle's and Bush's chief of staff, quickly got straight to the business at hand and told Quayle that Colin Powell was replacing him on the ticket.

The president had already agreed to the change, Quayle was told. "But there is an opportunity that you might want to be aware of," Skinner said, explaining that the next day the Republican nominee for governor in Indiana would be withdrawing from the race. "The state committee is hoping that you might be willing to bail them out of this tough spot by agreeing to run for the governor's seat," Skinner said, adding a patronizing, "It would be a good opportunity for you."

After the meeting, Quayle placed several phone calls, trying to gather support from conservatives who had been his strongest supporters. But Skinner had been on the telephone the day before, laying the groundwork for the change. Although all of his supporters wished him the best and offered future help in any political campaign, no one was willing to intercede on his behalf. After a night's sleep, Quayle decided to follow the plan, which he correctly assumed had been orchestrated by Baker, and announce that he would remove himself from the Bush ticket to attempt to become governor of Indiana.

As he watched Quayle's announcement on CNN, Baker couldn't resist a bit of gloating. "I've just signed Danny Quayle's death warrant," he laughed. "After he goes back to Indiana and loses to Bayh, ten years from now nobody will know who he is."

At campaign stops, Bush stuck to the story that Quayle had volunteered to leave the ticket and hadn't been pushed. After delivering a speech to the American Chamber of Commerce, Bush came the closest to allowing the story to crack in response to a critical question from a Quayle supporter. "You don't understand how difficult it is to make these decisions," Bush said. "It was Tension City in there. But you have to go forward. Dan

Quayle made his own decision—it was his, I didn't tell him to do it, he's his own man—but now we can go forward and so can he."

Bush-Powell turned out to be the dream team that the Republicans hoped that it would be. In the November election, Bush easily won reelection with 54 percent of the popular vote, to Clinton's 34 percent and Perot's 12 percent. But the upset of the election day was in Indiana, where Quayle surprised the pundits by defeating Bayh. This was the second time that Quayle had defeated a member of the Bayh family, having beaten Evan's father Birch Bayh for the Senate seat in 1980. In that election, the elder Bayh had reportedly dismissed his aides' suggestions that he needed to prepare for an upcoming debate, saying, "Come on, boys, don't bother me. I'm debating Danny Quayle." This victory was just as sweet for Quayle.

Quayle convinced William Kristol, who had been his chief of staff, to accompany him to Indiana for one year to get his administration off to a good start. Quayle, Kristol, and Quayle's lieutenant governor, Mitchell Daniels, a former Reagan White House political director, were able to assemble a staff of people from some of the brightest stars of conservative politics. As the national media checked in from time to time, expecting to write about Quayle's pratfall as the state's chief executive, they invariably returned to their editors empty-handed. Even Quayle's critics had to agree that the people in Indiana were pleased with the performance of their new governor.

As the 1996 election approached, Republicans assumed that Vice President Powell would be their nominee. But Powell was sincere in his dislike for politics, and he didn't have the voracious ambition required to become the nation's chief executive. In the middle of his term, Powell announced that he would not be a candidate for president under any circumstances.

A few conservative Quayle supporters from his days in the Bush administration (including Kristol, who had returned to Washington as planned) began mentioning Quayle as a possible choice, but, like Powell, Quayle took himself out of consideration in 1995. This choice was well received in Indiana, where the popular governor easily won reelection.

In 1996, the Republicans saw their sixteen-year hold on the presidency end when Democratic presidential nominee Albert Gore defeated Kansas senator and Republican nominee Robert Dole for the nation's highest office.

That meant that once again the Republican field for the presidency was wide open for the 2000 race. That spring, as the first primary was about to begin, the one person most Republican conservatives were pinning their hopes on was the once-ridiculed vice president, Dan Quayle.

Bush kept Quayle on the ticket in 1992, and no one will ever know whether that might have made a difference in the election. As for Quayle, he made a brief attempt at a run for the presidency in 1996 but exited quickly when he failed to gather any visible signs of support. He tried again in 2000 with the same result. In 1955, Eisenhower had urged Vice President Nixon to leave that office to do something more substantive in order to improve his political standing. Instead, Nixon served another term as veep and lost in 1960. If Quayle had taken

Eisenhower's advice to Nixon and left the administration after one term as vice president, he might have been taken more seriously as a presidential candidate. Unfortunately for Quayle, the lesson seemed to be that a politician can overcome anything except ridicule.

Chapter 28

The Conviction of Bill Clinton

President Bill Clinton was impeached for lying about his affair with a White House intern. What if the House of Representatives had viewed the affair as personal business and declined to impeach the president?

IF he's not telling the truth, his presidency is numbered in days," Sam Donaldson predicted on January 25, 1998. News about President Bill Clinton's extramarital affair with a young White House intern had come to light that week, and the commentariat of *ABC News* saw a quick finish to the story. "This thing is not going to drag out," Donaldson continued. "We're not going to be here months from now talking about this. Mr. Clinton—if he's not telling the truth and the evidence shows that—will resign, perhaps this week."

"His presidency, Cokie, is dead," added conservative pundit George Will.

"Okay, so he's out," said moderator Cokie Roberts. "So what happens next?"

"What will President Al Gore do then?" Donaldson asked.

"He'll select a very respected figure as vice president and will have a big honeymoon," answered William Kristol, talking head and editor of the conservative publication *The Weekly Standard*.

Bill Clinton wasn't out, and the commentators were in those same seats a year later *still* talking about the scandal. Clinton, America's most famous (and self-proclaimed) comeback kid, and the president who inspired more dirty jokes than any other chief executive in history, survived the scandal over the Lewinsky affair to finish out his term.

❧

Just after he won reelection, President Bill Clinton faced a political crisis. In the weeks before the 1996 presidential ballot, accusations first appeared that the Democratic National Committee had accepted at least $3 million from illegal donors, chiefly foreign donors and businesses. The DNC said that the donations were a series of errors and tried to end the matter by giving back the money. The next summer, a Senate Governmental Affairs Committee began looking into illegal campaign contributions, and more damaging information about the Democrats' fund-raising came to light.

There were several embarrassing revelations. The Clinton White House had offered a night in the Lincoln Bedroom to anyone who ponied up enough cash, and for those who found that price tag too steep, there was a series of coffees with the president for donors. Vice President Al Gore had attended a fund-raiser at a temple where Buddhist nuns had given large donations. This particular act violated several laws. For one, it is illegal to hold political fund-raisers at religious sites, and the nuns were straw donors who were merely passing along contributions for others who wished to remain anonymous, which is also illegal. In July 1997, Senator Fred Thompson began formal hearings on the charges, declaring in his opening statement that many of the illegal donations had actually been made by the Chinese Communist government in an attempt to influence the election and American foreign policy. Although many Democrats labeled Thompson's comments a partisan attack, two weeks later Democratic Connecticut senator Joseph Lieberman said that he looked at the evidence and agreed with Thompson that the Chinese government had tried to influence the 1996 presidential election.

The Senate committee held thirty-two days of hearings but was stymied by the unavailability of witnesses. Ten witnesses subpoenaed by the committee fled the country, and thirty-five invoked their Fifth Amendment right to refuse to testify to avoid self-incrimination.

Just as Clinton seemed to have safely slipped the noose of scandal, on January 17, 1998, an Internet gossip journalist, Matt Drudge, ran a story claiming that the president had carried on an affair with a White House intern. Although the next day other news outlets dismissed the story as crude gossip, by midweek it was appearing in every newspaper and on every televised news program. By the weekend a conventional wisdom had emerged (echoed on ABC's *This Week*) that the story probably wasn't true but that if it was, the president would soon resign.

It was as if a news bulletin had popped up on the screen during a CNN report on campaign finance, announcing, "We interrupt this program with an important, late-breaking episode of *Baywatch*." From that point on, the farce of the president's affair with the one-time intern forced any substantive political drama off the stage.

The accusation about the president had come to light during a deposition in the civil case of an Arkansas woman, Paula

Jones, who had accused Clinton of sexual harassment while he was governor. Clinton had denied the affair with the young woman named in the accusation, who the public would soon learn was named Monica Lewinsky. As the rumors about the affair continued to dominate the news, Clinton was forced to make a statement about the allegations. On Monday, January 26, after coaching from a Hollywood producer friend on how to deliver his lines, Clinton held a news conference. "I want you to listen to me. I'm going to say this again. I did not have sexual relations with that woman—Miss Lewinsky," the president said, shaking his finger at the camera and the American people. "I never told anybody to lie, not a single time. Never. And I need to go back to work for the American people."

But what Clinton didn't realize was that his paramour, Miss Lewinsky, had for months been telling the tales of the affair to a friend, Linda Tripp, who had begun taping the conversations when it appeared that she might be called to testify in the sexual harassment case. And those tapes were in the hands of special prosecutor Kenneth Starr, who had been investigating Bill and Hillary Clinton's involvement in White-water, a failed Arkansas land deal. Starr would soon refocus his investigation on the question of whether President Clinton had committed perjury by denying having had an affair with Monica Lewinsky during the Paula Jones deposition.

And what a remarkably tacky and sordid liaison it was. It had started in early November 1995, when the intern Lewinsky had shown the president her thong underwear. The president had then invited the young woman into a private study, where she began the affair while he telephoned a congressman. The two had a second such romantic encounter, and then a few weeks later, Lewinsky introduced herself to the president again at a New Year's Eve party because it had become obvious to her that he didn't know her name.

During a later encounter, according to Lewinsky's testimony, she asked him, "Is this just about sex . . . or do you have some interest in trying to get to know me as a person?" The president replied by saying that he enjoyed spending time with her and then "unzipped his pants and sort of exposed himself." As the affair cooled, Lewinsky began to pressure the president to help find her a job. "I don't want to have to work for this position. I just want it to be given to me," she had written him in a note, adding, "I am NOT someone's administrative/executive assistant." She included with the note an erotic postcard and her thoughts on education reform (higher pay for teachers, eliminate tenure).

Into the summer Clinton and his associates continued to deny that the affair had taken place, when suddenly rumors began circulating about a blue dress. After one of her visits to see the president, Lewinsky had kept a blue dress that was alleged to have physical proof of the affair, which the news media often referred to euphemistically as "DNA evidence." When it was discovered that Clinton had left physical evidence on Lewinsky's dress, a few Democratic defenders briefly tried to claim that there was an innocent reason for the physical evidence. "Democrats on the Hill . . . desperately suggest the DNA resulted from the President brushing up against Monica during one of their 'official' interactions," reported The *New York Times*. But other Democrats saw the situation differently. A reporter overheard Ohio Democrat James Traficant Jr. muttering in the halls of the Capitol, "If it's on the dress, he must confess."

On August 17, 1998, Clinton said in testimony before Starr's grand jury that he had had an "inappropriate intimate relationship" with Lewinsky, but that his earlier statements in the Jones deposition had been true. When the president was asked about the accuracy of his attorney's statement in an earlier deposition that "there is absolutely no sex of any kind" between Clinton and Lewinsky, Clinton replied, "It depends on what the meaning of the word 'is' is."

That evening he went on national television to acknowledge that there had been an "inappropriate relationship," but then he criticized the investigation, saying that he had a right to a private life outside the scrutiny of investigators.

Although many editorial boards around the country thought that the president's offenses were probably not impeachable, more than fifty major newspapers, including two of the nation's largest, *USA Today* and the *Wall Street Journal*, ran editorials urging President Clinton to resign.

Just a few weeks later, Starr completed his investigation of the Lewinsky affair, and in early September he delivered his report to the House of Representatives. Two days later, the Judiciary Committee voted to put the entire report, explicit language and all, onto the Internet for the public to view. The release of the Starr report on the Internet on a Friday afternoon brought business in the nation to a near standstill as millions of Americans logged on to read Starr's accusations. The report also touched off an impassioned debate over what should be done with Clinton. At times the rhetoric was overcooked: Author Toni Morrison wrote in *The New Yorker* that African Americans should support Bill Clinton—who had a Southern working-class background—because he was "our first black President."

There was a great debate in the nation over whether Bill Clinton's activities, those illegal and those immoral, were impeachable offenses. And for those faced with making exactly that decision, the choice was difficult. "I anticipate that I will conclude this matter the way I began it, somehow managing to irritate virtually everyone in my district who holds an opinion on the subject," said Representative Ed Pease, a Republican member of the Judiciary Committee who suffered a heart attack during the deliberations. "Those who believe there is nothing [to the charges] will be disappointed to know that I believe there is. Those who want me to do everything I can to vilify this president in every way possible will be disappointed to know my assessment of the facts cannot allow me to do that."

The 1996 midterm congressional elections were hailed by both Republicans and Democrats as a national referendum on the Clinton impeachment. The Republicans were expected to pick up only a few seats in each house if the election went poorly, and to pick up a large number of seats if the public was on their side. Instead, the Republicans lost seats in both houses. It was such a disastrous turn of events that the Republican Speaker of the House, Newt Gingrich, resigned from the House rather than face the embarrassment of being removed from the Speaker's seat by his Republican colleagues (it was later revealed during Gingrich's divorce that *he* had been carrying on an affair while presiding over the impeachment of Clinton).

After the election, the House Judiciary Committee began hearings on the Starr referral. The Starr report claimed that Clinton had obstructed justice, had committed perjury before a federal grand jury, and was guilty of being a no-class lout. The first two accusations were possibly impeachable offenses, and the House Judiciary Committee debated whether to issue articles of impeachment.

It sent a list of eighty questions to the president, beginning with one obviously designed to embarrass him: "Do you admit or deny that you are the chief law-enforcement officer of the United States of America?" Instead of providing a straightforward answer, Clinton, whom columnist Maureen Dowd had nicknamed "The Wizard of Is," submitted a long, rambling response purposefully written in a detached, passive style: "The President is frequently referred to as the chief law-enforcement officer, although nothing in the Constitution specifically designates the President as such. . . ."

In early December, after a week of hearings, the House Judiciary Committee passed four articles of impeachment against President Clinton for perjury, abuse of power, and obstruction of justice. "William Jefferson Clinton has under-mined the integrity of his office, has brought disrepute on the Presidency, has betrayed his trust as President, and has acted in a manner subversive of the rule of law and justice, to the mani-fest injury of the people of the United States," said Article One. "Wherefore William Jefferson Clinton, by such conduct, warrants impeachment and trial, and removal from office and disqualification to hold and enjoy any office of honor, trust or profit under the United States."

The articles of impeachment were sent to the full House, and the vote on whether to impeach President Clinton was scheduled for December 17.

The approaching vote on impeachment launched the most bizarre week in congressional history. On Wednesday, December 16, the eve of the vote, the United States launched Operation Desert Fox, an aerial bombardment of Iraq. The coincidence between the timing of the attack and the next day's vote seemed more than curious. Early Wednesday evening, Senate majority leader Trent Lott all but accused the president of beginning the attack simply to cause the House of Repre-sentatives to postpone the impeachment vote, although he quickly softened his comments in later interviews. Regardless, the bombing did cause the impeachment vote to be moved to Saturday.

On Thursday, Larry Flynt, the publisher of the porno-graphic magazine *Hustler*, announced that he had evidence that the House Speaker-elect, Bob Livingston of Louisiana, had not been faithful to his wife. Early Saturday morning, Republican leader Livingston began the impeachment proceedings with a call for the president to resign. As Democrats in the chamber began booing and yelling, "You resign," Livingston held up his hand to ask for quiet. As soon as order was restored, Livingston announced that he *was* going to resign, not just as Speaker but from his House seat as well. "I hope President Clinton will fol-low my example," Livingston said. (In a statement released that afternoon, Clinton declined Livingston's suggestion and offered his own suggestion that Livingston follow his example and keep his job.)

That afternoon, the Republican-controlled House voted largely along party lines to impeach Clinton.

Following the impeachment, Clinton gave an odd, tri-umphant speech declaring that he was a victim of "the politics of personal destruction." Vice President Al Gore, showing that sometimes being veep does mean being a lapdog, declared that the newly impeached president was a "man I believe will be regarded in the history books as one of our greatest presi-dents."

After the Christmas holidays, the articles of impeachment were submitted to the Senate, where they were received like a gift of moldy fruitcake. For weeks numerous senators had been saying that the Senate might not even hear the impeachment case, and when it became apparent that there might be a con-stitutional problem with that approach, the senators banded together to make sure that the trial was a trial in name only, refusing to hear witnesses or review evidence. Chris Canon, a Republican congressman from Utah, called the Senate reac-tion "a collective pant-wetting." The Senate trial opened with

Supreme Court chief justice William Rehnquist presiding as the judge. After two weeks of hearings, the Senate acquitted Clinton of the charges of perjury by a vote of fifty-five for conviction and forty-five against (an impeachment conviction requires a two-thirds majority). Thankful that the Senate had made it through the salacious trial without embarrassing itself, Senate majority leader Trent Lott was cheerful when Rehnquist's gavel fell, announcing the verdict. Lott thanked the chief justice at the end of the trial, saying, "Y'all come back soon."

WHAT IF?

What if Congress had decided not to impeach Clinton over his personal affair?

Every old pol knows how to count noses, and by early December, after three weeks of hearings, the seventy-four-year-old House leader Henry Hyde had a sense that he didn't have enough votes. Hyde, chairman of the House Judiciary Committee, was beginning to realize that the evidence special prosecutor Ken Starr had submitted to Congress wasn't everything the members of the committee had been expecting. Committee member Lindsey Graham, the conservative star from South Carolina, complained about the result to Hyde in the hallway. "He sent us over seven hundred pages, and there's nothing in there about Whitewater? Nothing about any of the other nonsense that's gone on? Listen, I have to go back home and go into churches and explain to folks what's happening and why. Does anybody expect me to take this soap opera and say this is why we're going to impeach the president? I don't know, Henry. I don't think this dog will hunt."

After the hearing Hyde assembled all of the Republican members of the committee into a room in the Capitol. As he did in each of their sessions, Hyde began by passing out cigars, not as a celebratory gesture, but as an inducement to conversation. As the congressmen lit their cigars (some more successfully than others), there was grumbling about the lack of evidence on other matters being investigated by Starr, and even about how the perjury over the sexual matter had been investigated. "I know he lied," Representative Cannon said about Clinton. "But this whole episode has the fragrance of entrapment. I've never spent a minute being sympathetic to Bill Clinton—I can't stand the man—but Starr's operation has the appearance of a perjury trap. Maybe you do this kind of sting to win a divorce case, but do you do this to impeach a president? The more I read it, the more uncomfortable I'm getting with the whole thing."

At the meeting, Hyde heard several of his fellow Republicans telling him the same thing: "I'm just too uncomfortable with it." And Hyde knew that the Democrats on the committee were almost gleeful at what they saw as Starr's tactical mistake. By including in his referral only complaints about the president's personal affairs, Starr made the referral too easy to dismiss as a lapse in personal behavior that didn't rise to the level of an impeachable offense.

Finally Hyde had heard enough. He stubbed out his cigar in a nearby ashtray, signalling to the other congressmen that the discussion was over. Hyde rose from his chair slowly. "When I came here, I thought I was going to change the world," he said. "Now I just hope that I can shuffle off with my dignity intact." Hyde took a deep breath and called his legislative aide, asking him to begin drafting the language of a motion to return the impeachment referral to the special prosecutor.

Hyde submitted the motion in the Judiciary Committee hearing the next day. John Conyers, the Democratic representative from Michigan who was the ranking minority member, interrupted Hyde before he could bring the motion to a vote. "Mr. Chairman, if you'll allow . . ."

"I'll yield."

"Mr. Chairman, I don't think we should conclude our business on this matter without taking a vote on impeachment or reaching some sort of verdict. Our president deserves to have some sort of closure on this."

"Well, I . . . uh," Hyde placed his hand over the microphone in front of him so that he could quietly consult with his Republican colleagues. "No, with all due respect to my esteemed colleague, I think that we should proceed with my motion to return this . . . ah . . . to send this referral back to Judge Starr until he has further evidence."

Hyde's motion passed the committee twenty-nine to eight. Six of the eight dissenting votes were from entrenched Republicans who were convinced that Clinton's affair with a White House intern was reason enough to impeach him whether he had committed any crimes or not, and the remaining two dissents came from Democratic members who apparently simply voted against the motion because a majority of the Republican committee members had favored it.

Thanks to the Judiciary Committee's unexpected Christmas gift, Bill Clinton was soon able to enjoy the happiest days of his presidency. Although he hadn't been cleared of any charges, he knew that the matter was would not raise its ugly head again. The nation's economy was the strongest it had ever been—the Dow had just cleared 10,000—and the nation was actually enjoying a budget surplus. He would soon be able to focus on more important issues, such as improving Social Security, that had been set aside for the past several months.

Not that Clinton's political enemies had given up. The aborted impeachment attempt only caused more money than ever to flow into the scandal-inspired slush funds. Likewise, in the newsrooms of the nation's major media outlets, all of the political reporters were racing for the next big Clinton story. Leads on Clinton scandals were available—sometimes it seemed to the reporters as if you could go down to the local Safeway and buy them in bunches of four for a dollar like green onions. But when news of the next Clinton scandal appeared, it wasn't from the right-wing political action groups or an investigative reporter, but from another congressional committee.

In January 1999, Congressman Christopher Cox released his committee's report on espionage by the Chinese government. The report claimed that the communist government in Beijing had stolen many of the United States' most important high-technology secrets, everything from nuclear warhead designs to the guidance system of the F-117 stealth fighter. Although the report said that the espionage had begun in the 1950s, when a scientist at the Jet Propulsion Laboratory delivered secrets about the Titan missile to China, the worst damage had occurred under President Clinton's watch.

As a result of the findings of the Cox committee, the Senate Governmental Affairs Committee reopened its earlier investigation of the fund-raising scandals to explore whether any connection existed between fund-raising in the 1996 presidential campaign and security breaches by China. From the beginning of the renewed hearings, Fred Thompson, the Tennessee senator chairing the committee, suspected there was a connection. "This deal's not any one real big thing," he said. "It's a lot of things strung together that paint a real ugly picture. Right now, we can put two and two together and we keep coming up with three and a half. But I think we're going to find that other half here somewhere."

The committee was assisted by Democratic fund-raiser Yah Lin "Charlie" Trie, who had pled guilty to felony charges of violating federal election laws, and by fund-raising Johnny Chung and John Huang. The stonewall that had stopped the committee in 1997 had developed a few holes.

Thompson's committee soon discovered that although President Clinton had informed congressional investigators that he had first learned of the Chinese espionage in 1998, he had actually been briefed about the security lapses in 1995—before the election, and before the money from China began flowing into his reelection campaign.

The scandal caught the attention of the public in a way that Thompson's earlier hearings never did. With the uncovering of the Chinese espionage, the matter extended beyond the dry subject of campaign finance, and the news organizations began covering the scandal as if it were the only story in the nation. MSNBC went for the new story full-bore. Each hour alternated between a correspondent standing in front of the White House or the Justice Department or some other Washington building, saying in a grave voice that although nothing much more was known, everyone there was taking the situation very seriously. Then the programs would switch to the pundits in the studio, and the producers would punch up the "Brady Bunch" shot, with four people all on-screen, all talking at the same time. If there was an intriguing idea offered in those debates, the viewers weren't able to hear it.

The Thompson committee was able to prove that many of the illegal contributions to the Democratic National Committee in 1996 had originated from the Chinese government. Chinese government officials used straw donors in the United States to pass along large donations. The donors had persuaded Clinton administration officials to approve the sale of many high-technology products, ranging from supercomputers to ultraprecise machining equipment, to nonmilitary companies in China. These companies, like the donors, were fake, set up merely to shift the equipment over to the Chinese military.

"I was never—I emphasize never—confronted at the time with any evidence or suggestion of willful misconduct, foreign government influence, sale of office, contributions in violation of the Federal Elections Campaign Act, or any other legal problems of any kind," testified Richard Sullivan, the boyish-looking former finance chief of the Democratic National Committee, who had been John Huang's supervisor.

"It appears, then, that you and the administration were duped," Thompson said.

"We were not duped, Senator."

"Well, you must have been duped somewhere along this road, or I'm guessing that you wouldn't have decided to give back $4 million in donations."

The committee found no evidence that anyone in the Clinton administration knew the source of the money at the time, but the committee's investigators found considerable evidence that the administration had tried to cover up the mess once they did learn of the China connection. As with most serious Washington scandals, it wasn't the crime but the cover-up that became the issue.

Although President Clinton had been chastised by Democratic senators over the Monica Lewinsky scandal, that was gentle teasing compared to the rhetoric coming from members of his own party over the apparent intertwining of Chinese espionage and illegal campaign contributions. Senator Joseph Biden appeared on *Fox News Sunday* and announced, "If there is any evidence that any political official in the White House or the State Department or anywhere in the administration knew that there was a correlation, it should be ferreted out, and that person should be indicted and put in jail."

"Do you mean *any* person?" prompted moderator Brit Hume.

"That person should be indicted no matter *who* it is," Biden repeated.

That same Sunday, Daniel Patrick Moynihan appeared on ABC's *This Week* program. In response to George Will's question about what should happen if evidence showed that the president was somehow involved, a sad-faced Moynihan said, "Let's get the truth out and punish those involved and be done with it, and then let's see that this doesn't ever happen again."

With Democratic senators making veiled allusions to impeachment, White House strategists decided they needed a big play to stanch the bleeding. They decided to burn Vice President Al Gore.

The vice president had earned the nickname of "solicitor in chief" during the 1996 campaign for his aggressive fund-raising style, but he had run into trouble over a fund-raiser held in April 1996 at a Buddhist temple in Los Angeles. Just as Richard Nixon had pushed along Spiro Agnew's investigation twenty-five years before, in a futile attempt to divert attention away from his own legal problems, White House aides began whispering to reporters that Al Gore was the real problem in the 1996 election.

When thirteen memos from White House aides to the vice president that referred to hard money fund-raising were leaked to the press, Gore said simply that he didn't read many memos sent to him, and hadn't read any of the memos in question. The press snickered at this explanation, but the giggles turned to guffaws when Gore said that he also hadn't heard of the illegal fund-raising activities in meetings because he spent quite a bit of time going to the restroom, explaining that he drank quite a bit of iced tea during the day.

None of this information was considered front page news, but then hundreds of e-mail messages between the White House and the vice president's office that were sent during 1994 to 1997—messages that had been considered lost—were released to congressional investigators. The e-mail messages made it clear that the vice president understood completely the nature of the fund-raising activities that he had denied knowing about, and suddenly the vice president was in serious trouble. The Sunday morning telepunditry was convinced that Gore would resign within days. On the ABC News program *This Week*, George Will, bow tie twitching with indignation, described Gore's explanation as "an absurdity wrapped in condescension" and insisted that if Gore had a shred of decency he would resign (although Cokie Roberts pointed out that Gore's iced-tea explanation reminded many longtime Washington watchers of the time that Vice President George Bush explained that he didn't know anything about the Iran-contra arms deal because "I didn't attend that meeting . . . I was at the Army-Navy game"). For several days the press speculated over whether Gore would become the third U.S. vice president to resign.

But Al Gore was smarter than Agnew, if the faint praise isn't too damning, and Gore's staff had damaging information of their own about the president's involvement and knowledge of what was happening with the Chinese campaign donations. Gore considered Clinton a personal friend, but he had noticed (as had the rest of the nation) that Clinton's friends, such as Webb Hubbell and Susan McDougal, sometimes went to prison to protect the president. Gore's friendship didn't extend to a federal prison, no matter how nice Clinton's aides were sure to make it sound. Within days, photocopies of memos that were under subpoena by the Senate committee, but that the White House staff had not been able to find, were being leaked to reporters. The memos indicated that many of the vice president's activities that were under investigation had been performed at the direction of the White House staff. Several of the memos had "BC" and a checkmark in the margins, showing that the president had read them.

In April, with the appearance of the memos, many of the former witnesses before Senator Thompson's committee filed amendments to their earlier testimony. The memos showed that in some donors' minds, at least, there was something of a connection between the technology transfers and the large donations, since often the donations had been delivered along with the requests for the relaxing of trade barriers. Because the Thompson committee had issued subpoenas in 1997 for all documents relating to the donations, the memos had clearly been withheld illegally.

In May 1999, the House Judiciary Committee began new impeachment hearings based on findings from the Cox report and Senator Thompson's committee. The Republicans drew up three articles of impeachment against the president: One article accused the president of perjury for the statements delivered to Congress concerning when he knew about the Chinese espionage. One alleged obstruction of justice for withholding documents from Senator Thompson's committee. And one alleged abuse of power for authorizing the waivers of the technology sales to reward campaign donors.

There was no doubt that the Judiciary Committee, which was controlled by the Republicans, would approve the three impeachment articles. The House of Representatives voted to impeach the president three weeks later, with nearly forty Democratic representatives joining the Republicans in condemning the president. Bill Clinton thus became the first president in 131 years to be impeached.

The impeachment then went to the Senate, where various senators speechified for five weeks. When it appeared that all one hundred were finally out of breath, on August 1, 1999, Supreme Court chief justice William Rehnquist began the historic roll call. "Senators, how say you?" Rehnquist asked after the entire text of each of the three articles was read aloud by the clerk. "Is the respondent, William Jefferson Clinton, guilty or not guilty?"

As the senators called our their verdicts some angry Republicans shouted "Guilty!" Some senators said "Not guilty" with as little emotion as possible, and a few appeared to be choking back tears as they announced their votes. Several senators were keeping their own tally on notepads, and when the sixty-seventh senator announced "Guilty!" the senators gasped. Minutes later, Rehnquist, who enjoyed the drama of the moment, pounded his gavel for effect. "Having been found guilty, William Jefferson Clinton will be removed from the Office of the President of the United States immediately."

In 1528 Sir Thomas Gresham, who would later become the financial agent of England's Queen Elizabeth I, noticed that whenever there were two coins of the same denomination, but one was made of a precious metal and the other was not, the bad money would soon be the only money on the market. Gresham's Law is the economic theory that says bad money chases good money off the market.

In 1998, America experienced the political equivalent of Gresham's Law. While the Thompson committee was investigating serious and substantive charges of Chinese influence in the 1996 presidential election, the energies of Congress, the Justice Department, and the media were directed instead at the Monica Lewinsky scandal. By the time the Cox committee's report on Chinese espionage in the United States was released in the spring of 1999—a report that did not show any links between the espionage and the illegal campaign contributions by the Chinese government—a sort of national scandal fatigue had set in. There was the general feeling that President Clinton had been punished enough for his sins by being impeached, and there was a general weariness that prevented anyone from wanting to hear about more scandals.

Monica Lewinsky had said in a conversation taped by Linda Tripp that she was tired of dealing with Bill Clinton. Lewinsky's comment accurately reflected how many people in the nation felt about presidential scandals by 1999: "If I ever want to have an affair with a married man again, especially if he's president, please shoot me."

Index